Universitext

Editors

F.W. Gehring
P.R. Halmos
C.C. Moore

Universitext

Editors: F.W. Gehring, P.R. Halmos, C.C. Moore

Booss/Bleecker: Topology and Analysis
Chern: Complex Manifolds Without Potential Theory
Chorin/Marsden: A Mathematical Introduction to Fluid Mechanics
Cohn: A Classical Invitation to Algebraic Numbers and Class Fields
Curtis: Matrix Groups, 2nd ed.
van Dalen: Logic and Structure
Devlin: Fundamentals of Contemporary Set Theory
Edwards: A Formal Background to Mathematics I a/b
Edwards: A Formal Background to Higher Mathematics II a/b
Endler: Valuation Theory
Frauenthal: Mathematical Modeling in Epidemiology
Gardiner: A First Course in Group Theory
Godbillon: Dynamical Systems on Surfaces
Greub: Multilinear Algebra
Hermes: Introduction to Mathematical Logic
Hurwitz/Kritikos: Lectures on Number Theory
Kelly/Matthews: The Non-Euclidean, The Hyperbolic Plane
Kostrikin: Introduction to Algebra
Luecking/Rubel: Complex Analysis: A Functional Analysis Approach
Lu: Singularity Theory and an Introduction to Catastrophe Theory
Marcus: Number Fields
Meyer: Essential Mathematics for Applied Fields
Moise: Introductory Problem Course in Analysis and Topology
Øksendal: Stochastic Differential Equations
Porter/Woods: Extensions of Hausdorff Spaces
Rees: Notes on Geometry
Reisel: Elementary Theory of Metric Spaces
Rey: Introduction to Robust and Quasi-Robust Statistical Methods
Rickart: Natural Function Algebras
Schreiber: Differential Forms
Smoryński: Self-Reference and Modal Logic
Stanisić: The Mathematical Theory of Turbulence
Stroock: An Introduction to the Theory of Large Deviations
Tolle: Optimization Methods

Lectures on Number Theory

Presented by
Adolf Hurwitz

Edited for Publication by
Nikolaos Kritikos

Translated, with some additional material, by
William C. Schulz

Springer Verlag
New York Berlin Heidelberg Tokyo

Nikolaos Kritikos
Parnithos 48
154 52 Psychiko
Athens
Greece

William Schulz (*Translator*)
Northern Arizona University
Department of Mathematics
Flagstaff, AZ 86011
U.S.A.

AMS Classifications: 10-01, 10A05, 10A15, 10A32, 10C99

Library of Congress Cataloging in Publication Data
Hurwitz, Adolf
   Lectures on number theory.
   (Universitext)
   Translated from the German.
   Bibliography: p.
   Includes index.
   1. Numbers, Theory of. I. Kritikos, Nikolaos,
1894–   . II. Title.
QA241.H85 1986   512.77   85-25093

Printed and bound by R.R. Donnelley & Sons, Harrisonburg, Virginia.
Printed in the United States of America.

9 8 7 6 5 4 3 2 1

ISBN 0-387-96236-0 Springer-Verlag New York Berlin Heidelberg Tokyo
ISBN 3-540-96236-0 Springer-Verlag Berlin Heidelberg New York Tokyo

Preface

During the academic year 1916-1917 I had the good fortune to be a student of the great mathematician and distinguished teacher Adolf Hurwitz, and to attend his lectures on the Theory of Functions at the Polytechnic Institute of Zürich. After his death in 1919 there fell into my hands a set of notes on the Theory of numbers, which he had delivered at the Polytechnic Institute. This set of notes I revised and gave to Mrs. Ferentinou-Nicolacopoulou with a request that she read it and make relevant observations. This she did willingly and effectively. I now take advantage of these few lines to express to her my warmest thanks.

Athens, November 1984

N. Kritikos

About the Authors

ADOLF HURWITZ was born in 1859 at Hildesheim, Germany, where he attended the Gymnasium. He studied Mathematics at the Munich Technical University and at the University of Berlin, where he took courses from Kummer, Weierstrass and Kronecker. Taking his Ph.D. under Felix Klein in Leipzig in 1880 with a thesis on modular functions, he became Privatdozent at Göttingen in 1882 and became an extraordinary Professor at the University of Königsberg, where he became acquainted with D. Hilbert and H. Minkowski, who remained lifelong friends. He was at Königsberg until 1892 when he accepted Frobenius' chair at the Polytechnic Institute in Zürich (E.T.H.) where he remained the rest of his life.

Hurwitz's mathematics was heavily influenced by Felix Klein. He worked mainly in number theory and related areas of complex analysis, including modular functions, Riemann surfaces, and complex multiplication. Hurwitz originated the invariant volume for integration on the orthogonal groups, which was later generalized to Haar measure on topological groups. He showed that the real numbers, complex numbers, quaternions and Cayley octaves are the only algebras without divisors of zero and with quadratic norm over the real numbers. This result became one of the pillars of the theory of algebras. Hurwitz did pioneering work on the arithmetic of quaternions, and discovered the most fruitful definition of an "integral" quaternion. Also interesting are his papers on various aspects of continued fractions.

Hurwitz died in Zürich in 1919.

NIKOLAOS KRITIKOS was born of Greek parents in Constantinople in 1894. He studied Mathematics at the Universities of Athens, Göttingen and Zürich, and at the Polytechnic Institute (E.T.H.) in Zürich. Among his teachers were C. Caratheodory and Adolf Hurwitz. He received his Dr. Phil. from the Philosophical Faculty II of the University of Zürich in 1920 with a dissertation written under Dr. G. Polya. He became full Professor of Higher Mathematics at the University of Thessaloniki in 1928. During the years 1933-1946 and 1951-1963 he served as full Professor of Higher Mathematics at the E.M. Polytechnic Institute in Athens.

Translator's Preface

This English version of A. Hurwitz's Lectures on Number Theory has been taken from the edited version of Prof. N. Kritikos, with occasional consultation of the original notes. A very few modifications have been incorporated into the last chapter of the English version to take advantage of the greater familiarity of present day students with matrices.

The translator wishes to point out the splendid organization used by Prof. Hurwitz. For example, in the last chapter the theory of binary quadratic forms and the theory of continued fractions are developed together in about the same space which would be necessary to develop the theory of continued fractions alone. Another example is the formula (40.1) derived from Gauss' lemma, from which follows the law of quadratic reciprocity and both complementary theorems.

In order to make the book more useful as a classroom text, the translator has added problems at the end of the chapters. These are of three types. The first, numerical examples, have been constructed with the aim of providing insight into the general situation with the least amount of calculation. The second class of problems provides computational algorithms for the theoretical material covered in the book. The third class of problems attempts to provide interesting extensions of the theory in the main text.

The main text is completely independent of the problems. Problems are not organized by degree of difficulty. Rather, an attempt has been made to roughly correlate the problems with the sections of the text, so that they may be worked as the chapter is read. The problems are for the most part easy, and copious hints have been provided, so that they may be solved in a reasonable time.

The translator would like to thank Prof. Kritikos for his active cooperation in the correction of the English text. He would also like to thank Evelyn Wong and Kim Poole of the Ralph M. Bilby Research Center for their fine job of typing, and Northern Arizona University for its support of the entire project, and my wife Maria M. Schulz for her generous contributions of time and effort on the project.

The translator would greatly appreciate it if any errors detected by readers are forwarded to him, as well as suggestions for additional problems or improvement in the existing ones.

William C. Schulz
Mathematics Department
Northern Arizona University
Flagstaff, Arizona USA 86011

Table of Contents

CHAPTER 1. BASIC CONCEPTS AND PROPOSITIONS

1. The Principle of Descent..1

2. Divisibility and the Division Algorithm..............................3

3. Prime Numbers...6

4. Analysis of a Composite Number into a Product of Primes..............8

5. Divisors of a Natural Number n, Perfect Numbers....................12

6. Common Divisors and Common Multiples of two or more Natural Numbers..15

7. An Alternate Foundation of the Theory of
        The Greatest Common Divisor....................................18

8. Euclidean Algorithm for the G.C.D. of two Natural Numbers...........21

9. Relatively Prime Natural Numbers...................................23

10. Applications of the Preceding Theorems............................26

11. The Function $\phi(n)$ of Euler....................................32

12. Distribution of the Prime Numbers in the Sequence
        of Natural Numbers..37

        Problems for Chapter 1..45

CHAPTER 2. CONGRUENCES

13. The Concept of Congruence and Basic Properties....................51

14. Criteria of Divisibility...53

15. Further Theorems on Congruences..................................56

16. Residue Classes mod m...58

17. The Theorem of Fermat...60

18. Generalized Theorem of Fermat...................................61

19. Euler's Proof of the Generalized Theorem of Fermat...................62

    Problems for Chapter 2...66

CHAPTER 3. LINEAR CONGRUENCES

20. The Linear Congruence and its Solution..............................68

21. Systems of Linear Congruences......................................71

22. The Case when the Moduli m_1, m_2,...,m_k of the System of

    Congruences are pairwise Relatively Prime..........................74

23. Decomposition of a Fraction into a Sum of

    An Integer and Partial Fractions...................................76

24. Solution of Linear Congruences with the aid

    of Continued Fractions...79

    Problems for Chapter 3...84

CHAPTER 4. CONGRUENCES OF HIGHER DEGREE

25. Generalities for Congruence of Degree $k > 1$ and Study

    of the Case of a Prime Modulus.....................................89

26. Theorem of Wilson...93

27. The System $\{r, r^2, ..., r^\delta\}$ of Incongruent Powers Modulo a prime p......94

28. Indices...96

29. Binomial Congruences..99

30. Residues of Powers Mod p..101

31. Periodic Decadic Expansions.....................................103

    Problems for Chapter 4...106

CHAPTER 5. QUADRATIC RESIDUES

32. Quadratic Residues Modulo m...109

33. Criterion of Euler and the Legendre Symbol..........................109

34. Study of the Congruence $x^2 \equiv a \pmod{p^r}$....................112

35. Study of the Congruence $x^2 \equiv a \pmod{2^k}$...................116

36. Study of the Congruence $x^2 \equiv a \pmod m$ with $(a,m)=1$......120

37. Generalization of the Theorem of Wilson............................123

38. Treatment of the Second Problem of §32.............................127

39. Study of $(\frac{-1}{p})$ and Applications.........................128

40. The Lemma of Gauss...129

41. Study of $(\frac{2}{p})$ and an application........................133

42. The Law of Quadratic Reciprocity..................................135

43. Determination of the Odd Primes p for which $(\frac{q}{p}) = 1$ with given q...138

44. Generalization of the Symbol $(\frac{a}{p})$ of Legendre by Jacobi.............139

45. Completion of the Solution of the Second Problem of §32.............146

Problems for Chapter 5...151

CHAPTER 6. BINARY QUADRATIC FORMS

46. Basic Notions...157

47. Auxiliary Algebraic Forms...160

48. Linear Transformation of the Quadratic Form $ax^2 + 2bxy + cy^2$........161

49. Substitutions and Computation with them............................162

50. Unimodular Transformations (or Unimodular Substitutions)............168

51. Equivalence of Quadratic Forms....................................170

52. Substitutions Parallel to $\begin{pmatrix} 0 & -1 \\ 1 & 0 \end{pmatrix}$ 172

53. Reductions of the First Basic Problem of §46.......................174

54. Reduced Quadratic Forms with Discriminant $\Delta < 0$.....................178

55. The Number of Classes of Equivalent Forms with Discriminant $\Delta < 0$...184

56. The Roots of a Quadratic Form.......................................187

57. The Equation of Fermat (and of Pell and Lagrange)...................192

58. The Divisors of a Quadratic Form...................................198

59. Equivalence of a form with itself and solution of the Equation
    of Fermat for Forms with Negative Discriminant Δ..................201

60. The Primitive Representations of an odd Integer by $x^2 + y^2$.........203

61. The Representation of an Integer m by a Complete System
    of Forms with given Discriminant $\Delta < 0$...........................205

62. Regular Continued Fractions.......................................213

63. Equivalence of Real Irrational Numbers............................219

64. Reduced Quadratic Forms with Discriminant $\Delta > 0$....................226

65. The Period of a Reduced Quadratic Form With $\Delta > 0$...................232

66. Development of $\sqrt{\Delta}$ in a Continued Fraction.........................241

67. Equivalence of a form with itself and solution of the equation
    of Fermat for forms with Positive Discriminant Δ..................243

    Problems for Chapter 6...252

BIBLIOGRAPHY...265

INDEX...272

Chapter 1
Basic Concepts and Propositions

1. THE PRINCIPLE OF DESCENT

A basic property of the integers

$$0, \pm1, \pm2, \pm3 ,\ldots, \pm n, \pm(n+1), \ldots \tag{1.1}$$

is the following:

The sum $a+b$, the difference $a-b$ and the product ab of two integers is again an integer.

We express this property by saying that addition, subtraction and multiplication may be carried out without restriction within the set (1.1) of integers.

The positive integers

$$1, 2, 3, \ldots, n, n+1, \ldots \tag{1.2}$$

are called natural numbers. The natural numbers have the following basic property:

Each non-empty set Φ of natural numbers contains a least element; in other words there exists an element a_e in Φ satisfying the relation

$$a_e \leqslant x \quad \text{for every} \ x \in \Phi. \tag{1.3}$$

In fact, by the hypothesis there exists at least one element a in Φ; then the first element a_e in the series $1, 2, 3, \ldots, a$ which is a member of Φ has the property (1.3). For example in the set $\mathbb{N}$ of natural numbers the least element is 1, and in the set of natural numbers which are sums of at least two (equal or unequal) natural numbers the least element is $2 = 1+1$.

A DESCENDING SEQUENCE OF NATURAL NUMBERS is a sequence of such numbers

$$a_1 > a_2 > a_3 > \ldots \tag{1.4}$$

1.1.2

For every such sequence we have the following basic proposition:

<u>PROPOSITION</u> Every descending sequence of natural numbers terminates, that is, it has a final, least, element, and hence consists of a finite number of natural nmbers.

In fact, if the first element of the sequence is a_1, then the number of elements in the sequence is at most equal to a_1.

We will call the above proposition the PRINCIPLE OF DESCENT. From it results the following method of proof which has been used frequently in Number Theory since the epoch of the great French number theorist Pierre de Fermat (1601-1665):

If a hypothesis has as a consequence the formation of a non-terminating descending sequence of natural numbers, then this hypothesis must be rejected.

1.2.1

2. DIVISIBILITY AND THE DIVISION ALGORITHM

An integer α is called divisible by an integer β if there exists an integer γ such that $\alpha = \beta\gamma$. The integer β is then called a divisor of α and this relationship is symbolized by

$$\beta \mid \alpha.$$

Simple consequences of this definition are

1. Each integer α is a divisor of itself; for $\alpha = \alpha \cdot 1$

2. The integer 1 is a divisor of any integer; for $\alpha = 1 \cdot \alpha$

3. The integer 0 is divisible by any integer α ; for $0 = \alpha \cdot 0$

4. The integer 0 divides only itself; for $\alpha = 0 \cdot \gamma$ implies that $\alpha = 0$.

We easily find that

$$(\alpha \mid \beta \text{ and } \beta \mid \gamma) \text{ imply } \alpha \mid \gamma.$$

Actually, from the hypothesis we have

$$\beta = \alpha \cdot \alpha_1 \text{ and } \gamma = \beta \cdot \beta_1$$

with integers α_1 and β_1; so

$$\gamma = \alpha \cdot \alpha_1 \beta_1$$

with integer $\alpha_1 \beta_1$.

THEOREM (Division Algorithm) If n is a natural number and a is any integer, then the following equality holds:

$$a = qn + r \quad \text{where q and r are integers and } 0 \leq r < n. \tag{2.2}$$

This representation of a is uniquely determined, that is, q and r are uniquely defined integers.

<u>Proof</u> For $a = 0$, (2.2) obviously holds with $q = r = 0$. Let now $a > 0$. We form the set M of those natural numbers which are integral multiples xn of n and exceed a: $x \cdot n > a$. The set M is not empty, since, for example, when $x >$ a then

$$xn > an \geqslant a.$$

Hence M has a least element, which we may designate by $(q+1)n$ with integer $q \geqslant 0$. Then

$$q \cdot n \leqslant a < (q+1)n,$$

so

$$0 \leqslant a-qn = r < (q+1)n - qn = n$$

and thus

$$a = qn + r \quad \text{with} \quad 0 \leqslant r < n.$$

Finally, let $a < 0$. Then the relation

$$-a = q'n + r' \quad \text{with } 0 \leqslant r' < n$$

will hold. If $r' = 0$ then

$$a = -q'n = qn + r \quad \text{with } q = -q' \text{ and } r = 0.$$

If $0 < r' < n$ then

$$a = -q'n - r' = (-q'-1)n + n-r' = qn + r$$

$$\text{with } q = -q'-1 \text{ and } 0 < r = n-r' < n.$$

Now we will show that q and r are uniquely determined in the representation (2.2) of a. Let $a = qn + r$ and $a = q'n + r'$ with q, q', r, r' integers and $0 \leqslant r, r' < n$. Then the following will hold

$$0 = (q-q')n + r-r'$$

and hence

$$r'-r = (q-q')n.$$

But $r'-r$ is less than n in absolute value, $|r'-r| < n$, while $(q-q')n$ is either 0 (when $q = q'$) or at least equal to n in absolute value (when $q \neq q'$). Hence for $|q-q'|n = |r-r'|$ it is necessary and sufficient that $|q-q'|n = |r-r'| = 0$, that is, that $r = r'$ and $q = q'$, which is just what was to be proved.

1.2.3

Before advancing with the study of the properties of divisibility, we note that if $\alpha = \beta\gamma$ where α, β, γ are integers, then also $|\alpha| = |\beta||\gamma|$ where $|\alpha|$, $|\beta|$, $|\gamma|$ are integers greater than or equal to 0. Hence if α is divisible by β then $|\alpha|$ is divisible by $|\beta|$. On the other hand, if $|\alpha| = |\beta||\gamma|$, then $\alpha = \beta \cdot (\pm\gamma)$, so that, if $|\alpha|$ is divisible by $|\beta|$ then α is divisible by β. Thus the study of divisibility in the set of integers can be reduced to the study of divisibility in the set of non-negative integers. Hence in the remaining portion of this chapter we will deal with the non-negative integers.

1.3.1

3. PRIME NUMBERS

A natural number $n > 1$ with only two positive divisors is called prime.

Examples of prime numbers are 2, 3, 13, 101.

<u>THEOREM</u> 1 Let a be a natural number, $a > 1$ and a not a prime. Then the smallest divisor of a different from 1 is a prime number.

<u>Proof</u> Since a is not prime, there is at least one divisor of a which is different from 1 and a. Let p be the least among the divisors of a different from 1 and a. Then

$$a = pt \quad \text{with} \quad 2 \leqslant p \leqslant a-1 \quad \text{and} \quad p \leqslant t \leqslant a-1$$

since $t \mid a$ and $t \neq 1, a$. If p were not prime, we would have

$$p = p_1 t_1 \quad \text{with} \quad 1 < p_1 < p \; ;$$

hence p_1 would be a divisor of a different from 1 and a and smaller than p. Consequently p would not be the smallest among the divisors of a different from 1 and a. Thus the smallest divisor p must be a prime number, and from the relation $a = pt$ with $p \leqslant t$ follows $a \geqslant p \cdot p = p^2$ so $p \leqslant \sqrt{a}$.

<u>Corollary</u> Every natural number $a > 1$ has at least one prime divisor.

The proof is immediate.

<u>THEOREM</u> 2 There are infinitely many primes.

<u>Proof</u> In fact, if p_1, p_2, ..., p_k are prime numbers, then the number $(p_1 p_2 \ldots p_k) + 1$ either is prime and obviously larger than each of p_1, p_2, ..., p_k or is not prime but has a prime divisor q. This divisor q cannot be equal to any of p_1, p_2, ..., p_k because in the case $q = p_j$ $(1 \leqslant j \leqslant k)$ we would have

$$q \mid p_1 p_2 \ldots p_k \quad \text{and} \quad q \mid (p_1 p_2 \ldots p_k + 1)$$

and thus

$$q \mid (p_1 p_2 \ldots p_k + 1) - p_1 p_2 \ldots p_k = 1 \; ;$$

1.3.2

but this is impossible since, as a prime, $q \geqslant 2$. Thus the primes $p_1, p_2, \ldots, p_k$ do not exhaust the set of prime numbers, however large the natural number k. Hence we conclude that the set of prime numbers is not finite.

4. ANALYSIS OF A COMPOSITE NUMBER INTO A PRODUCT OF PRIMES

If the integer $a > 1$ is not prime, then it is composite. For composite numbers we have the following proposition:

THEOREM 1 Each composite number is a product of primes.

Proof Let $a > 1$ be composite; according to Theorem 1 of §3 we will have $a = pa_1$, where p is a prime number and a_1 an integer with $1 < a_1 < a$. If a_1 is prime then a is a product of two primes and the theorem is verified. If a_1 is composite, then we will have $a_1 = p_1 a_2$ with p_1 prime and $1 < a_2 < a_1 < a$. If a_2 is prime then $a = pp_1 a_2$ and the theorem is proved. If a_2 is composite, then the above procedure may be continued and leads to an equation $a = pp_1 p_2 a_3$ with p_2 prime and $1 < a_3 < a_2 < a_1 < a$. According to the principle of descent, the descending sequence of natural numbers $a > a_1 > a_2 > a_3 > \ldots$ must terminate with some a_k such that $a = pp_1 p_2 \ldots p_{k-1} a_k$ and this necessitates that a_k be prime. Hence a is equal to a product of $k + 1$ primes.

THEOREM 2 The decomposition of a composite number into a product of primes is unique, if we overlook the order of the factors.

We will base the proof on a corollary of the following lemma.

LEMMA If p is a prime and $1 \leqslant a < p$, $1 \leqslant b < p$, then p is not a divisor of the product ab; $p \nmid ab$.

Proof From the hypotheses $a < p$ and $b < p$ it follows that $p \nmid a$ and $p \nmid b$. Therefore if $a = 1$ or $b = 1$ then $p \nmid ab$ and the claim of the theorem is verified. Thus let $a > 1$ and $b > 1$. We will suppose that $p \mid ab$ and we will show that this hypothesis has as a consequence the formation of a nonterminating descending sequence of natural numbers and hence must be rejected.

According to the theorem of §2 there exist q and a_1 satisfying

1.4.2

$$p = aq + a_1 \text{ with } q \geqslant 1 \text{ and } 0 < a_1 < a. \tag{4.1}$$

Indeed $q \geqslant 1$ since the dividend p is greater than the divisor a, and the remainder $a_1 > 0$ since if a_1 were 0 it would imply that the prime p would be divisible by a with $a > 1$ and $a < p$, which is impossible. From the relation (4.1) it follows that

$$pb = abq + a_1 b$$

so

$$a_1 b = pb - abq = pb - pmq \text{ with integer } m$$

since $p|ab$.

Therefore $a_1 b$ is divisible by p, so we have

$$a_1 b = pm_1 \text{ with } 0 < a_1 < a < p.$$

We may now repeat the above procedure and obtain the relation

$$p = a_1 q_1 + a_2 \text{ with } 1 \leqslant q_1 \text{ and } 0 < a_2 < a_1.$$

From this we conclude once more that

$$a_2 b = pb - a_1 b q_1 = p(b - m_1 q_1)$$

so $p|a_2 b$. The procedure may now be repeated without end, and leads to a nonterminating descending sequence of natural numbers

$$a > a_1 > a_2 > \ldots. \text{ with } p|a_k b.$$

Consequently the hypothesis $p|ab$ must be rejected and thus $p \nmid ab$.

From this lemma we may immediately derive the following two corollaries.

<u>Corollary 1*</u> If p is a prime, $p \nmid a$ and $p \nmid b$ then $p \nmid ab$.

  <u>Proof</u> In fact we have

$$a = pq + a_1 \text{ with } q \geqslant 0 \text{ and } 0 < a_1 < p$$
$$b = ps + b_1 \text{ with } s \geqslant 0 \text{ and } 0 < b_1 < p$$

---

* This is now often called Euclids Lemma.

so

$$ab = p^2 qs + pb_1 q + pa_1 s + a_1 b_1$$
$$= p\,(pqs + b_1 q + a_1 s) + a_1 b_1 \quad \text{with } 0 < a_1 b_1.$$

According to the lemma, $p \nmid a_1 b_1$; therefore we have $p \nmid ab$ since the hypothesis $p \mid ab$ together with $p \mid p(pqs + b_1 q + a_1 s)$ would imply $p \mid a_1 b_1$.

<u>Corollary 2</u> If $p \mid abc...m$ where p is prime and a, b, ..., m are natural numbers, then p divides at least one of the factors a, b, ..., m.

<u>Proof</u> According to Corollary 1, it is not possible simultaneously to have $p \nmid a$ and $p \nmid bc...m$. If $p \mid a$ then the assertion of the Corollary is verified. If $p \nmid a$ then we have $p \mid bc...m$. Thus we will have either $p \mid b$ or $p \nmid b$ and $p \mid (c...m)$. Continuing in the same way we will find a factor of the product $abc...m$ which is divisible by p.

Naturally, the proof of Corollary two could be given the form of mathematical induction as follows:

According to Corollary 1, Corollary 2 is valid for a product of two factors. Let us suppose that Corollary 2 is valid for any product of $k \geqslant 2$ factors; we will show it valid for any product of $k + 1$ factors. In fact, if $p \mid a_1 a_2 ... a_k a_{k+1}$ then $p \mid (a_1 a_2 ... a_k) a_{k+1}$. Thus, by Corollary 1, either $p \mid a_1 ... a_k$ and then according to the induction hypothesis p divides at least one of the factors a_1, a_2, ..., a_k; or $p \nmid a_1 ... a_k$ and thus $p \mid a_{k+1}$. Thus Corollary 2 is verified for a product of $k + 1$ factors.

Now we may prove Theorem 2 which constitutes a basic proposition of Number Theory. Let

$$n = p_1 \cdots p_r = q_1 \cdots q_s \quad \text{with } r \leqslant s$$

be two decompositions of the natural number n into products of primes. The prime number p_1 is a divisor of n and thus of the product $q_1 ... q_s$. Hence,

1.4.4

according to Corollary 2, p_1 is a divisor of at least one of q_1, q_2, $\ldots q_s$. Let $p_1 | q_j$ where $1 \leqslant j \leqslant s$. But q_j is a prime and has only 1 and q_j as divisors. Since $p_1 \neq 1$, we have $p_1 = q_j$, and thus, dividing out p_1,

$$p_2 \; \cdots \; p_r = q_1 \; \cdots \; q_{j-1} q_{j+1} \; \cdots \; q_s.$$

Likewise we find that p_2 is identical with at least one of q_1, $\ldots$, q_{j-1}, q_{j+1}, $\ldots$, q_s; p_3 with some other factor q of the product and finally we arrive at the relation

$$p_r = q_1' \; \cdots \; q_{s-r+1}'$$

where $q_1' \; \cdots \; q_{s-r+1}'$ are some remaining factors of the decomposition of n as a product of q_1, $\ldots$, q_s. Since p_r is a prime, this relation implies first that $s-r = 0$, that is $s = r$, and second that $p_r = q_1'$. Thus the theorem has been proved.

<u>Remark</u> In the decomposition of n as a product $p_1 p_2 \cdots p_r$ of primes, it is possible that the factors p may be equal in groups. If we gather these equal factors together, we find the following expression for n:

$$n = p_1^{\alpha_1} p_2^{\alpha_2} \cdots p_k^{\alpha_k}$$

where $k \geqslant 1$ is a natural number, $p_i \neq p_j$ for $i \neq j$ and α_1, α_2, $\ldots$, α_k are natural numbers. Moreover, this expression for n is unique if we label the prime factors of n, when $k \geqslant 2$, in such a way that $p_1 < p_2 < \ldots < p_k$. This unique expression we will call the decomposition of the natural number n into prime factors or the representation of n by prime factors.

Henceforth when we write $n = p_1^{\alpha_1} p_2^{\alpha_2} \cdots p_k^{\alpha_k}$ without accompanying conditions for p and α, it will mean this particular representation of the natural number n by prime factors.

1.5.1

5. DIVISORS OF A NATURAL NUMBER n. PERFECT NUMBERS.

If $n = 1$ then the only divisor of n which is a natural number is 1. If n = p is a prime then there are two natural number divisors of n: 1 and p.

Let $n > 1$ be composite. Then

$$n = p_1^{\alpha_1} p_2^{\alpha_2} \ldots p_k^{\alpha_k}$$

with $k \geqslant 1$, $p_i \neq p_j$ if $k \geqslant 2$ and $i \neq j$, and α_1, α_2, ..., α_k natural numbers. If $k = 1$ then $\alpha_1 > 1$. We have

<u>THEOREM</u> 1 For the natural number d to be a divisor of $n = p_1^{\alpha_1} p_2^{\alpha_2} \ldots p_k^{\alpha_k}$ it is necessary and sufficient for d to have the form

$$d = p_1^{\beta_1} p_2^{\beta_2} \ldots p_k^{\beta_k} \quad \text{with } 0 \leqslant \beta_j \leqslant \alpha_j \text{ for } j = 1, \ldots, k.$$

<u>Proof</u> Every prime divisor of d will be a divisor of n so it cannot be distinct from p_1, p_2, ..., p_k. Hence d must be of the form

$$d = p_1^{\beta_1} p_2^{\beta_2} \ldots p_k^{\beta_k} \quad \text{where } 0 \leqslant \beta_i \leqslant \alpha_i \quad \text{for } j = 1, \ldots, k.$$

On the other hand, a natural number d of the above form is a divisor of n, because

$$n = (p_1^{\beta_1} p_2^{\beta_2} \ldots p_k^{\beta_k})(p_1^{\alpha_1 - \beta_1} p_2^{\alpha_2 - \beta_2} \ldots p_k^{\alpha_k - \beta_k})$$

<u>THEOREM</u> 2 The number of natural number divisors of $n = p_1^{\alpha_1} p_2^{\alpha_2} \ldots p_k^{\alpha_k}$ is

$$(\alpha_1 + 1)(\alpha_2 + 1) \ldots (\alpha_k + 1).$$

For example, the number of divisors of $48 = 2^4 \cdot 3^1$ is $(4+1)(1+1) = 10$. Indeed, the divisors are 1, 2, 3, 4, 6, 8, 12, 16, 24, 48.

<u>THEOREM</u> 3 The sum $Z(n)$ of the natural number divisors of n is

$$Z(n) = \sum_{d|n} d = (1 + p_1 + p_1^2 + \ldots + p_1^{\alpha_1}) \ldots (1 + p_k + p_k^2 + \ldots + p_k^{\alpha_k})$$

$$= \frac{p_1^{\alpha_1 + 1} - 1}{p_1 - 1} \ldots \frac{p_k^{\alpha_k + 1} - 1}{p_k - 1}.$$

The proof is immediate, since on multiplying out we get 1 and all the products of primes which are divisors of n.

For example, the sum of the divisors of $48 = 2^4 \cdot 3^1$ is

$$\frac{2^5-1}{2-1} \cdot \frac{3^2-1}{3-1} = 31 \cdot 4 = 124.$$

THEOREM 4 If $n = p^\alpha \cdot m$ where p does not divide m, then

$$Z(n) = \frac{p^{\alpha+1}-1}{p-1} \cdot Z(m) = Z(p^\alpha)\, Z(m)$$

The proof is immediate.

DEFINITION A natural number n is _perfect_ if it is the sum of its natural divisors which are less than itself.

Since the only natural divisor of n excluded in the definition is n itself, n is perfect if and only if $Z(n) = n+n = 2n$. An even natural number is of the form

$$n = 2^{\lambda-1}u \quad \text{where } \lambda \geqslant 2 \text{ and } u \text{ is odd.}$$

According to Theorem 4 we have

$$Z(n) = (2^\lambda-1)\, Z(u).$$

If n is perfect, then $Z(n) = 2n$. Hence

$$(2^\lambda-1)\, Z(u) = 2n = 2^\lambda \cdot u$$

so

$$Z(u) = \frac{2^\lambda u}{2^\lambda-1} = \frac{(2^\lambda-1)u+u}{2^\lambda-1} = u + \frac{u}{2^\lambda-1}.$$

From this follows that $u/(2^\lambda-1)$ is an integer and consequently $2^\lambda-1$ is a divisor of u. Hence the quotient $u/(2^\lambda-1)$ is also a divisor of u.

Since $\lambda > 2$, we have now found that

$$Z(u) = u + (\text{a divisor of } u \text{ which is smaller than } u).$$

Hence the sum $Z(u)$ of the divisors of u is equal to u itself plus a second divisor of u which is smaller than u. This implies that u has only two divisors, and the second divisor must be 1, so u is prime and $u/(2^{\lambda}-1) = 1$ and consequently $u = 2^{\lambda}-1$. So we have finally found that if the even number n is perfect then it has the form $n = 2^{\lambda-1}(2^{\lambda}-1)$ with $\lambda > 2$ and $2^{\lambda}-1$ a prime number.

On the other hand if $n = 2^{\lambda-1}(2^{\lambda}-1)$ with $\lambda > 2$ and $2^{\lambda}-1$ a prime number then (according to Theorem 4 and since $2^{\lambda}-1$ is prime)

$$Z(n) = \frac{2^{\lambda}-1}{2-1} \cdot Z(2^{\lambda}-1) = (2^{\lambda}-1)(1+2^{\lambda}-1) = 2^{\lambda}(2^{\lambda}-1)$$

$$= 2 \cdot 2^{\lambda-1}(2^{\lambda}-1) = 2n,$$

so n is a perfect number.

For example $6 = 2^{2-1}(2^2-1) = 2 \cdot 3$ and $28 = 2^{3-1}(2^3-1) = 4 \cdot 7$ and $496 = 2^{5-1}(2^5-1) = 16 \cdot 31$ are perfect.

<u>Remark</u> For $2^{\lambda}-1$ to be a prime it is necessary that λ be prime. Actually, if this is not the case and $\lambda = mn$ with $1 < m, n < \lambda$ then $2^{\lambda}-1 = 2^{mn}-1$ is not prime because

$$2^{mn}-1 = (2^m)^n-1 = (2^m-1)(2^{m(n-1)}+2^{m(n-2)}+ \ldots +2^m+1).$$

The condition that λ be prime in order that $2^{\lambda}-1$ be prime is necessary but not sufficient, because, for example, $2^{11}-1 = 2047 = 23 \cdot 89$. On the other hand $2^{13}-1 = 8191$ is prime. Consequently

$$2^{12} \cdot (2^{13}-1) = 4096 \cdot 8191 = 33,550,336$$

is a perfect number.

6. COMMON DIVISORS AND COMMON MULTIPLES OF TWO OR MORE NATURAL NUMBERS.

We consider first the common divisors of two natural numbers a and b, excluding the trivial special case $a = b = 1$. Evidently we may then write

$$a = p_1^{\alpha_1} p_2^{\alpha_2} \cdots p_k^{\alpha_k}$$

$$b = p_1^{\beta_1} p_2^{\beta_2} \cdots p_k^{\beta_k}$$

where $k \geqslant 1$, p_1, p_2, ..., p_k are prime numbers different from one another, (if $k \geqslant 2$), and $\alpha_j \geqslant 0$, $\beta_j \geqslant 0$ and $\alpha_j + \beta_j > 0$ for $j = 1, \ldots, k$. (For example, if $a = 28$ and $b = 1$ we have $p_1 = 2$, $p_2 = 7$, $\alpha_1 = 2$, $\alpha_2 = 1$ and $\beta_1 = 0$, $\beta_2 = 0$.) The common divisors of a and b are obviously the natural numbers

$$d = p_1^{\gamma_1} p_2^{\gamma_2} \cdots p_k^{\gamma_k} \quad \text{where } 0 \leqslant \gamma_j \leqslant \text{Min } \{\alpha_j, \beta_j\} = m_j, \ j = 1, \ldots, k.$$

Therefore there exists a greatest common divisor

$$(a,b) = p_1^{m_1} p_2^{m_2} \cdots p_k^{m_k}.$$

We will denote the greatest common divisor of a and b by (a,b).

THEOREM 1   Each common divisor of a and b is a divisor of the greatest common divisor (a,b), and conversely each divisor of the g.c.d. (a,b) is a common divisor of a and b.

The proof is immediate.

DEFINITION   Two or more numbers are called _relatively_ _prime_ if their only positive common divisor is 1.

THEOREM 2 $a = p_1^{\alpha_1} \cdots p_k^{\alpha_k}$ and $b = p_1^{\beta_1} \cdots p_k^{\beta_k}$ are relatively prime if and only if $m_j = \text{Min } \{\alpha_j, \beta_j\} = 0$ for $j = 1, \ldots, k$.

The proof is immediate.

1.6.2

<u>THEOREM</u> 3 If $(a,b) = d$ and $a = da'$ and $b = db'$ then $(a',b') = 1$.

  <u>Proof</u> We have, with the above symbolism,

$$a' = p_1^{\alpha_1-m_1} \ldots p_k^{\alpha_k-m_k}, \quad b' = p_1^{\beta_1-m_1} \ldots p_k^{\beta_k-m_k}$$

and

$$(a',b') = p_1^{Min\{\alpha_1-m_1,\ \beta_1-m_1\}} \ldots p_k^{Min\{\alpha_k-m_k,\ \beta_k-m_k\}}.$$

However, we have the equalities

$$Min\ \{\alpha_j-m_j,\ \beta_j-m_j\} = 0 \quad for\ j = 1,\ \ldots,\ k,$$

since at least one of the two numbers $\alpha_j-m_j \geqslant 0$ and $\beta_j-m_j \geqslant 0$ is equal to 0 for $j = 1,\ \ldots,\ k$.

<u>Comment</u> On this last proposition is based the determination of partial fractions equal to a given fraction a/b.

  A positive common multiple of the numbers $a = p_1^{\alpha_1}\ldots p_k^{\alpha_k}$ and $b = p_1^{\beta_1}\ldots p_k^{\beta_k}$ is one of the infinitely many numbers

$$m = p_1^{\mu_1} \ldots p_k^{\mu_k} \cdot u \quad where\ \mu_j \geqslant Max\ (\alpha_j,\ \beta_j) = M_j\ for\ j = 1,\ \ldots,\ k\ \ and$$

u is a natural number.

  There exists among all the common multiples a least, the least common multiple (lcm)

$$p_1^{M_1} p_2^{M_2} \ldots p_k^{M_k}.$$

<u>THEOREM</u> 4 Each common multiple of a,b is divisible by the least common multiple, that is, is a multiple of the least common multiple. Conversely, every multiple of the least common multiple of a and b is a common multiple of a and b.

  The proof is immediate.

<u>THEOREM</u> 5 For two natural numbers a and b

$$(greatest\ common\ divisor\ of\ a,b) \cdot (least\ common\ multiple\ of\ a,b) = ab.$$

1.6.3

<u>Proof</u>

$$(\text{GCD of } a,b)\cdot(\text{LCM of } a,b) = p_1^{m_1+M_1} \ldots p_k^{m_k+M_k}$$

and

$$m_j+M_j = \text{Min } \{\alpha_j, \beta_j\} + \text{Max } \{\alpha_j, \beta_j\} = \alpha_j + \beta_j \quad \text{for } j = 1, \ldots, k$$

so

$$(\text{GCD of } a,b)\cdot(\text{LCM of } a,b) = p_1^{\alpha_1+\beta_1} \ldots p_k^{\alpha_k+\beta_k} = ab.$$

The case of three or more natural numbers we may study in the same way. For example, if we have three natural numbers

$$a = p_1^{\alpha_1} \ldots p_k^{\alpha_k}, \quad b = p_1^{\beta_1} \ldots p_k^{\beta_k}, \quad c = p_1^{\gamma_1} \ldots p_k^{\gamma_k}$$

with $\alpha_j, \beta_j, \gamma_j \geqslant 0$ and $\alpha_j + \beta_j + \gamma_j > 0$ for $j = 1, \ldots, k$ and we put

$$m_j = \text{Min } \{\alpha_j, \beta_j, \gamma_j\}, \quad M_j = \text{Max } \{\alpha_j, \beta_j, \gamma_j\} \quad (j = 1, \ldots, k)$$

then the common divisors of a,b,c are the numbers of the form

$$p_1^{a_1} \ldots p_k^{a_k} \quad \text{where } 0 \leqslant a_j \leqslant m_j \qquad j = 1, \ldots, k$$

and the common multiples are the numbers of the form

$$p_1^{b_1} \ldots p_k^{b_k}\cdot u \quad \text{where } M_j \leqslant b_j \ (j = 1, \ldots, k) \text{ and } u \text{ is a natural}$$

number.

Therefore there exists a greatest common divisor of a,b,c, the natural number

$$p_1^{m_1} \ldots p_k^{m_k}$$

and a least common multiple, the natural number

$$p_1^{M_1} \ldots p_k^{M_k}.$$

Theorems 1 to 4 continue to hold also in this case, after, of course, some obvious changes in their phraseology.

Similarly, we define three numbers to be relatively prime when their greatest common divisor is 1. They are relatively prime if and only if there exists no prime which is a common divisor of all three.

7. AN ALTERNATE FOUNDATION OF THE THEORY OF THE GREATEST COMMON DIVISOR.

<u>DEFINITION</u> A non-empty set of complex numbers is called a module* if it is closed under addition and subtraction; that is, the sum and difference of two elements of the set is an element of the set.

For example, the set $\{0\}$ is a module, as is the set $\{0, \pm2, \pm4, \pm6, \ldots\}$ of even integers. The module $\{0\}$ is the only module with a finite number of members. Actually, a module containing a number $a \neq 0$ will also contain all the infinitely many numbers $2a = a+a$, $3a = 2a+a$, $\ldots$, na, $\ldots$.

<u>THEOREM</u> 1 If the (complex) numbers a_1, a_2, $\ldots$, a_k belong to a module M, then M also contains all numbers of the form $\sum_{j=1}^{k} x_j a_j$, where the x_j are integers.

Proof M contains the element 0, since $0 = a_1 - a_1$. The numbers $-a_1$, $-a_2$, $\ldots$, $-a_k$ belong to M since $-a_j = 0 - a_j$, $j=1$, $\ldots$, k. Thus along with the a_j, any number of the form $x_j a_j$ with x_j an integer belongs to M. So the sum $\sum_{j=1}^{k} x_j a_j$ also belongs to M.

<u>THEOREM</u> 2 Let a_1, $\ldots$, a_k be k (complex) numbers. Then the set of all numbers of the form $\sum_{j=1}^{k} x_j a_j$ where $x_j \in Z = $ (set of all integers), forms a module.

This module will be symbolized by $[a_1, \ldots, a_k]$ and is called the module <u>generated</u> by a_1, $\ldots$, a_k.

Proof Actually, if
$$a = \sum_{j=1}^{k} x_j' a_j \quad \text{and} \quad b = \sum_{j=1}^{k} x_j'' a_j \quad \text{with } x_j', x_j'' \text{ integers}$$
then also the numbers
$$a + b = \sum_{j=1}^{k} (x_j' + x_j'') a_j \quad \text{and} \quad a - b = \sum_{j=1}^{k} (x_j' - x_j'') a_j$$
are elements of the set of numbers of the form $\sum_{j=1}^{k} x_j a_j$.

* Strictly speaking, a module over the integers, but this is the only type of module we will consider.

<u>THEOREM</u> 3 If M is a module which contains a real number a ≠ 0, then it will contain an infinite number of positive numbers and an infinite number of negative numbers.

   <u>Proof</u> M will contain along with a also -a, so it will contain all numbers of the form ±na where n is a natural number. Among these numbers there will be infinitely many positive and infinitely many negative numbers.

<u>THEOREM</u> 4 If a module M ≠ {0} consists of integers, then it coincides with the set of integer multiples of some natural number.

   <u>Proof</u> M contains, as a consequence of Theorem 3, infinitely many positive integers. Among them there exists, as discussed in §1, a least positive integer d.

   Let now a be any integer in M. For it we have

$$a = dq + r \quad \text{with integer } q \quad \text{and} \quad 0 \leqslant r < d.$$

I say $r = 0$; indeed, if $r > 0$ then we would have

$$r = a - dq \in M$$

and the natural number r not only belongs to M but is smaller than d in contradiction to the defining property of d, that it be the least positive integer belonging to M.

   So for each $a \in M$ we have $a = qd$. On the other hand xd with integer x belongs to M. Thus

$$M = \{xd \mid x \in \mathbf{Z}\}.$$

<u>Corollary</u> Let $a_1, \ldots, a_k$ be integers not all 0. Then $[a_1, \ldots, a_k] = [d]$ for some natural number d.

The proof is immediate.

   <u>Remark</u> From $[a_1, \ldots, a_k] = [d]$ with $d > 0$ it follows that

$$a_j = dm_j \text{ with integer } m_j \ (j=1, \ldots, k).$$

1.7.3

Thus d is a common divisor of $a_1, \ldots, a_k$. However, since d is an element of the module $[a_1, \ldots, a_k]$, we have

$$d = g_1 a_1 + \ldots + g_k a_k \text{ with integers } g_1, \ldots, g_k.$$

Consequently, if an integer δ is a common divisor of $a_1, \ldots, a_k$, then $\delta \neq 0$ and δ divides d. That is, every common divisor of $a_1, \ldots, a_k$ is also a divisor of d. Therefore the natural number d is identical with the greatest common divisor of the natural numbers $|a_j|$ as it was defined in the previous section. For these reasons we will call the d defined by the above conditions the greatest common divisor of the integers $a_1, \ldots, a_k$.

8. EUCLIDEAN ALGORITHM FOR THE GREATEST COMMON DIVISOR (GCD) OF TWO NATURAL NUMBERS

Let a_0 and a_1 be two natural numbers with $a_0 \geqslant a_1$. The case $a_0 = a_1$ gives $a_0 = a_1 \cdot 1$ and thus a_1 is the greatest common divisor of a_0 and a_1. Let now $a_0 > a_1$. We apply repeatedly the division algorithm (§2)

$$a_0 = a_1 q_1 + a_2 \quad \text{with } q_1 \geqslant 1 \quad \text{and} \quad 0 \leqslant a_2 < a_1.$$

If $a_2 = 0$ then obviously

$$[a_0, a_1] = [a_1]$$

and a_1 is the greatest common divisor of a_0 and a_1.

Let now $a_2 \neq 0$. Then $[a_0, a_1] = [a_1, a_2]$. Actually, each $x_0 a_0 + x_1 a_1$ with integers x_0, x_1 is equal to

$$x_0 (a_1 q_1 + a_2) + x_1 a_1 = (x_0 q_1 + x_1) a_1 + x_0 a_2 \in [a_1, a_2].$$

Conversely, each $y_1 a_1 + y_2 a_2$ with integers y_1, y_2 is equal to

$$y_1 a_1 + y_2 (a_0 - q_1 a_1) = (y_1 - y_2 q_1) a_1 + y_2 a_0 \in [a_0, a_1].$$

Thus $[a_0, a_1] = [a_1, a_2]$.

We continue on in the same way. Since $a_2 \neq 0$ we will have

$$a_1 = a_2 q_2 + a_3 \quad \text{with} \quad q_2 \text{ an integer and } 0 \leqslant a_3 < a_2.$$

If $a_3 = 0$, then $[a_1, a_2] = [a_2]$ and thus

$$a_2 = GCD (a_1, a_2) = GCD (a_0, a_1)$$

since $[a_0, a_1] = [a_1, a_2]$.

If $a_3 \neq 0$, the process of division may be continued, and gives

$$a_2 = a_3 q_3 + a_4 \quad \text{with integer } q_3 \text{ and } 0 \leqslant a_4 < a_3.$$

Whenever the remainder of such a division is not equal to 0, the process may be continued, generating a descending sequence of natural numbers

$$a_0 > a_1 > a_2 > \ldots \; .$$

By the principle of descent, this sequence must terminate. Let this occur at the kth division, so that

$$a_k = a_{k+1} \, q_{k+1} + a_{k+2} \quad \text{with integer } q_{k+1} \text{ and } a_{k+2} = 0.$$

We will then have

$$[a_0,a_1] = [a_1,a_2] = [a_2,a_3] = [a_3,a_4] = \ldots = [a_k,a_{k+1}] = [a_{k+1}]$$

from which we conclude that $a_{k+1} = d = GCD\,(a_0,a_1)$.

The above series of divisions is called the Euclidean Algorithm. As we saw, the series terminates on appearance of a division with remainder 0. The divisor of this last division is the greatest common divisor of the originally given two natural numbers for which we sought.

1.9.1

9. RELATIVELY PRIME NATURAL NUMBERS

As we have seen, two natural numbers a and b are called relatively prime to one another, when the greatest common divisor (a,b) is equal to 1. For such numbers, where by numbers we mean natural numbers, the following hold:

__THEOREM__ 1 If each of the numbers a_1, ..., a_k are relatively prime to each of the numbers b_1, ..., b_ℓ then the product $a_1...a_k$ is relatively prime to the product $b_1...b_\ell$.

__Proof__ Let us suppose that the products $a_1...a_k$ and $b_1...b_\ell$ are not relatively prime to one another. There would then exist a prime p which is a divisor of both $a_1...a_k$ and $b_1...b_\ell$. The prime p as a divisor of $a_1...a_k$ would then, according to corollary 2 of §4, be a divisor of at least one a_i $(1 \leq i \leq k)$ of the numbers a_1, ..., a_k and as a divisor of $b_1...b_\ell$ would be a divisor of at least one b_j $(1 \leq j \leq \ell)$ of the numbers b_1, ..., b_ℓ. Thus p would be a common divisor of a_i and b_j and thus these two numbers would not be relatively prime, contrary to the hypothesis of the theorem.

__Corollary__ If (a,b) = 1 then $(a^r, b^s) = 1$ for $r \geq 0$ and $s \geq 0$.

The proof is immediate.

__DEFINITION__ The natural numbers a_1, a_2, ..., a_k are called pairwise relatively prime if for any two elements a_i and a_j, $i \neq j$, we have $(a_i,a_j) = 1$.

__THEOREM__ 2 If a_1, ..., a_k are pairwise relatively prime and if each of them divides the number b then their product $a_1...a_k$ divides b.

__Proof__ The theorem is obvious if $a_1 = a_2 = ... = a_k = 1$. Now let at least one of a_1, ..., a_k be larger than 1, so that $a_1 ... a_k > 1$, and let $a_1...a_k = p_1^{r_1} ... p_\ell^{r_\ell}$ be the decomposition of $a_1...a_k$ into a product of primes. The number $p_j^{r_j}$ will be a divisor of exactly one a_i $(1 \leq i \leq k)$ of the numbers a_1, ..., a_k since they are pairwise relatively prime. From $p_j^{r_j} | a_i$ and $a_i | b$ it

1.9.2

follows that $p_j^{r_j} | b$, and since this is true for $j = 1, \ldots, \ell$ we will have $p_1^{r_1} \ldots p_\ell^{r_\ell} | b$ and hence $a_1 \ldots a_k | b$.

<u>THEOREM</u> 3 If the product ab of two numbers a and b is divisible by a number γ and γ is relatively prime to a then γ is a divisor of b.

  <u>Proof</u> The numbers a and γ are, by hypothesis, relatively prime. Each of them is a divisor of ab. Hence by Theorem 2, we have $a\gamma | ab$, that is, $ab = a\gamma m$ with integer m. Thus $b = \gamma m$, that is $\gamma | b$.

<u>LEMMA</u> If $a = p_1^{\alpha_1} \ldots p_k^{\alpha_k}$ is equal to the nth power b^n of a number b, then the exponents $\alpha_1, \ldots, \alpha_k$ are multiples of n.

  <u>Proof</u> From the hypothesis $b^n = p_1^{\alpha_1} \ldots p_k^{\alpha_k}$ it follows that b has no prime factors distinct from $p_1, \ldots, p_k$.

Let

$$b = p_1^{\rho_1} \ldots p_k^{\rho_k}$$

be the decomposition of b into prime factors.

Then we will have

$$b^n = p_1^{\rho_1 n} \ldots p_k^{\rho_k n} = a = p_1^{\alpha_1} \ldots p_k^{\alpha_k}.$$

Applying Theorem 2 of §4 which shows the uniqueness of decomposition into prime factors, we have

$$\rho_1 n = \alpha_1, \ldots, \rho_k n = \alpha_k$$

which proves the assertion of the lemma.

An immediate consequence of this lemma is that for a natural number a to be equal to the nth power of a natural number b it is necessary and sufficient that in the decomposition $a = p_1^{\alpha_1} \ldots p_k^{\alpha_k}$ of a into prime factors the exponents α_j are multiples $n\sigma_j$ of n for $j = 1, \ldots, k$.

__THEOREM__ 4 If $a_1 \ldots a_k = b^n$ and the $a_1, \ldots, a_k$ are pairwise relatively prime then $a_j = A_j^n$ with integer A_j for $j=1, \ldots, k$.

__Proof__ Let p^{ρ_j} be the highest power of a prime p which is a divisor of a_j $(1 \leqslant j \leqslant k)$. This same power will appear in the decomposition of b^n into prime factors, since $p \nmid a_i$ for $i \neq j$. In agreement with the above we will have $\rho_j = n\sigma_j$ where σ_j is a natural number. So, in the decomposition of a_j into prime factors, the exponents of the factors will be multiples of n and hence a_j will be equal to the nth power of some integer A_j.

10. APPLICATIONS OF THE PRECEDING THEOREMS

FIRST APPLICATION. PYTHAGOREAN TRIPLES

Three natural numbers which satisfy the equation

$$x^2 + y^2 = z^2$$

are said to form a Pythagorean triple. For example, $x = 4$, $y = 3$, $z = 5$ is a Pythagorean triple. In other words, a Pythagorean triple is any solution (x_0, y_0, z_0) in natural numbers of the Diophantine equation $x^2 + y^2 = z^2$.

If $x_0^2 + y_0^2 = z_0^2$ and if d is the greatest common divisor of x_0, y_0, z_0, then we will also have

$$\left(\frac{x_0}{d}\right)^2 + \left(\frac{y_0}{d}\right)^2 = \left(\frac{z_0}{d}\right)^2 ,$$

and so $(x_0/d, y_0/d, z_0/d)$ is also a Pythagorean triple with the additional property that the three numbers which compose it are relatively prime, that is, have greatest common divisor 1. Such Pythagorean triples are called primitive. For example, a primitive triple is $(4, 3, 5)$. From it we may form infinitely many non-primitive triples $(4d, 3d, 5d)$ where $d \geq 2$ is a natural number.

DETERMINATION OF ALL PRIMITIVE PYTHAGOREAN TRIPLES

Let x, y, z be a primitive Pythagorean triple, that is

$$x^2 + y^2 = z^2 \quad \text{and} \quad (\text{GCD of } x, y, z) = 1.$$

The integers x, y must be relatively prime, for if a prime p were a common divisor, it would divide $x^2 + y^2$ and thus also z^2, and then z, and finally p would divide x, y, z, so $(x, y, z) > 1$ contrary to our hypothesis. Similarly x, z and y, z must be relatively prime pairs. However, the integers x, y cannot both be odd, since otherwise

$$(2k+1)^2 + (2\lambda+1)^2 = 4(k^2 + \lambda^2 + k + \lambda) + 2 = z^2$$

which is excluded because the square of any integer when divided by 4 leaves remainder 0 or 1, never 2. Hence we may assume without loss of generality that x is even and y is odd. Then z will be an odd integer.

From $x^2 + y^2 = z^2$ we conclude that

$$x^2 = z^2 - y^2 = (z+y)(z-y).$$

Since z and y are odd, $z+y$ and $z-y$ are even. Thus we will have

$$\frac{x^2}{4} = \frac{z+y}{2} \cdot \frac{z-y}{2} \quad \text{with} \quad \frac{z+y}{2}, \frac{z-y}{2} \text{ integers.} \tag{10.1}$$

We note that the integers $(z+y)/2$ and $(z-y)/2$ must be relatively prime, for if they had a common divisor p, p would also be a common divisor of

$$\frac{z+y}{2} + \frac{z-y}{2} = z \quad \text{and} \quad \frac{z+y}{2} - \frac{z-y}{2} = y$$

which was excluded above. Applying Theorem 4 of §9 to (10.1), we have

$$\frac{z+y}{2} = u^2, \quad \frac{z-y}{2} = v^2 \quad \text{with natural numbers } u \text{ and } v. \tag{10.2}$$

Thus we will have

$$z = u^2 + v^2 \quad \text{and} \quad y = u^2 - v^2. \tag{10.3}$$

Moreover, from (10.1) and (10.2) it follows that

$$\frac{x^2}{4} = u^2 v^2 \quad \text{so} \quad \frac{x}{2} = uv$$

and hence

$$x = 2uv \qquad\qquad (10.4)$$

To sum up, we have found that, if (x,y,z) is a primitive Pythagorean triple, with x even and y odd, then there exist two natural numbers u and v such that (10.3) and (10.4) hold. For the numbers u and v we conclude

$$(u,v) \leqslant (u^2,v^2) = (\tfrac{z+y}{2}, \tfrac{z-y}{2}) = 1$$

so $(u,v) = 1$. Moreover, it is not possible for both u and v to be odd, for in that case $y = u^2-v^2$ would be even while, as we have shown, y is odd.

On the other hand, let $u_1 > v_1$ be two natural numbers which are relatively prime $((u_1,v_1) = 1)$ and not both odd. Then the triple of natural numbers

$$x_1 = 2u_1v_1 \; , \; y_1 = u_1^2 - v_1^2 \; , \; z_1 = u_1^2 + v_1^2$$

satisfies the equation $x^2 + y^2 = z^2$, since

$$x_1^2 + y_1^2 = 4u_1^2v_1^2 + (u_1^2-v_1^2)^2 = (u_1^2+v_1^2)^2 = z_1^2.$$

Moreover, we have, since $(a,b)|(a+b,a-b)$,

$$(y_1,z_1) = (u_1^2-v_1^2, \; u_1^2+v_1^2)|(2u_1^2, \; 2v_1^2) = 2$$

so $(u_1^2-v_1^2, \; u_1^2+v_1^2) = 1$ or 2. But it is impossible that $(u_1^2-v_1^2, \; u_1^2+v_1^2) = 2$ since $u_1^2 - v_1^2$ and $u_1^2 + v_1^2$ are both odd, because one of u_1,v_1 is odd and the other even. Hence $(y_1,z_1) = 1$, and the triple (x_1,y_1,z_1) is primitive.

EXAMPLES

1. $u = 2, \; v = 1;$ $x = 4, \quad y = 3, \quad z = 5$

2. $u = 4, \; v = 1;$ $x = 8, \quad y = 15, \quad z = 17$

3. $u = 3, \; v = 2;$ $x = 12, \quad y = 5, \quad z = 13$

4. $u = 5, \; v = 2;$ $x = 20, \quad y = 21, \quad z = 29.$

1.10.4

SECOND APPLICATION

The French mathematician P. Fermat asserted that he had proved the proposition:

The Diophantine equation $x^n + y^n = z^n$ has no solutions in the set $\mathbb{N}$ of natural numbers for $n = 3, 4, 5, \ldots$.

Such a proof was not found among Fermat's papers for which reason the proposition is called the conjecture of Fermat, and it remains unproved. It is easy to see that the conjecture holds in the set $\mathbb{N}$ if it holds in the set $\mathbb{Z} - \{0\}$ of non-zero integers. Actually if n is an even integer then $x^n + y^n = z^n$ is equivalent to $(\pm x)^n + (\pm y)^n = (\pm z)^n$. If n is an odd integer and $x < 0$, $y > 0$, $z > 0$ then $x^n + y^n = z^n$ and $y^n = z^n + (-x)^n$. If n is odd and $x < 0$, $y > 0$, $z < 0$ then $x^n + y^n = z^n$ and $y^n + (-z)^n = (-x)^n$. Finally, if n is odd and $x < 0$, $y < 0$, $z < 0$ then $x^n + y^n = z^n$ and $(-x)^n + (-y)^n = (-z)^n$. The conjecture, for the set $\mathbb{N}$ of natural numbers, has up to now been proved for many values of $n \geq 3$, but not for all. We will now present an elementary proof for $n = 4$ which we owe to the German mathematician Kummer.

Since $x^4 + y^4 = z^4$ is equivalent to $x^4 + y^4 = (z^2)^2$, it suffices to prove the unsolvability of $x^4 + y^4 = z^2$ in the set $\mathbb{N}$ of natural numbers. We will show that the hypothesis of the existence of a solution leads to an infinite descending sequence of natural numbers, which is impossible by the principle of descent, and so the hypothesis must be rejected.

We will call the product $h = xyz$ the height of the solution. Let x_0, y_0, z_0 be a solution in natural numbers with height $h_0 = x_0 y_0 z_0$. We wish to determine a solution with smaller height. If $(x_0, y_0) \neq 1$, let p be a common prime divisor of x_0 and y_0, and thus p^2 a divisor of z_0; setting

$$x_1 = \frac{x_0}{p}, \qquad y_1 = \frac{y_0}{p}, \qquad z_1 = \frac{z_0}{p^2}$$

we have

$$p^4 x_1^4 + p^4 y_1^4 = p^4 z_1^2$$

so

$$x_1^4 + y_1^4 = z_1^4$$

and

$$h_1 = x_1 y_1 z_1 = \frac{x_0}{p} \frac{y_0}{p} \frac{z_0}{p^2} = \frac{h_0}{p^4} < h_0 \ .$$

If $(x_0,y_0) = 1$ then $(x_0^2,y_0^2) = 1$ (Corollary of Theorem 1 of §9) and thus the (GCD of x_0^2,y_0^2,z_0) = 1. Moreover $(x_0^2)^2 + (y_0^2)^2 = z_0^2$ so x_0^2, y_0^2, z_0 is a primitive Pythagorean triple. We then have, assuming without loss of generality that x_0^2 is even (whence y_0^2 and z_0 are odd)

$$x_0^2 = 2u_0 v_0, \quad y_0^2 = u_0^2 - v_0^2, \quad z_0 = u_0^2 + v_0^2 \tag{10.5}$$

with $u_0 > v_0 > 0$ and $(u_0,v_0) = 1$ and $2|u_0 v_0$.
Then we have

$$v_0^2 + y_0^2 = u_0^2.$$

Since y_0^2 is odd, v_0^2 must be even, and since (GCD of v_0,y_0,u_0) = 1 we again have a primitive Pythagorean triple v_0,y_0,u_0 with v_0 even and y_0 odd. There will then exist two natural numbers r and s with

$$v_0 = 2rs, \quad y_0 = r^2 - s^2, \quad u_0 = r^2 + s^2$$

with $r > s$ and $(r,s) = 1$ and $2|rs$. Applying these values to the first of the equation (10.5) we come to

$$x_0^2 = 2(r^2+s^2) \cdot 2rs = 4(r^2+s^2)rs,$$

so

$$(\tfrac{x_0}{2})^2 = (r^2+s^2)rs \tag{10.6}$$

But the three numbers $r^2 + s^2$, r, s are relatively prime in pairs. Indeed we know that $(r,s) = 1$ and

$$(r, r^2+s^2) \leqslant (r^2, r^2+s^2) = (r^2, r^2+s^2-r^2) = (r^2, s^2) = 1;$$

consequently $(r, r^2+s^2) = 1$. Similarly we find that $(s, r^2+s^2) = 1$. Now since r, s, r^2+s^2 are relatively prime in pairs, from (10.6) by Theorem 4 of §9 it follows that r, s, r^2+s^2 are squares of natural numbers:

$$r = x_1^2, \quad s = y_1^2, \quad r^2+s^2 = z_1^2.$$

Since

$$x_1^4 + y_1^4 = r^2+s^2 = z_1^2,$$

the triple x_1, y_1, z_1 is a new solution in natural numbers of $x^4 + y^4 = z^2$, with height

$$h_1 = x_1 y_1 z_1 = \sqrt{x_1^2 y_1^2 z_1^2} = \sqrt{rs(r^2+s^2)} = \sqrt{\left(\frac{x_0}{2}\right)^2} = \frac{x_0}{2} < x_0 y_0 z_0 = h_0.$$

Since repeating the procedure would lead to an infinite descending sequence of natural numbers

$$h_0 > h_1 > h_2 > \ldots,$$

the hypothesis that $x^4 + y^4 = z^2$ has a solution in $\mathbb{N}$ must be rejected.

11. THE FUNCTION $\phi(n)$ OF EULER

Let n be a natural number. Among the integers 1, 2, ..., n, that is, the positive integers $\leq$ n, there exist numbers which are relatively prime to n, for example, 1. The number of integers in the set {1, 2, ..., n} which are relatively prime to n is $\phi(n)$, the value at n of Euler's function ϕ. The domain of definition of ϕ is the set $\mathbb{N}$ of natural numbers. For example

$$\phi(1) = 1, \ \phi(2) = 1, \ \phi(3) = 2, \ \phi(4) = 2, \ \phi(5) = 4, \ \phi(6) = 2$$

and

$$\phi(p) = p-1$$

for each prime number p.

DETERMINATION OF THE FORMULA FOR THE VALUE $\phi(n)$ OF THE FUNCTION ϕ

Let $n = p_1^{\alpha_1}...p_k^{\alpha_k}$ with $\alpha_j \geq 1$ (j=1, ..., k) be the decomposition of n into prime factors. We have to find the number of positive integers $\leq$ n which are not divisible by p_1, nor by p_2, ..., nor by p_k. We note that the numbers $\leq$ n which are divisible by p_1 are p_1, $2p_1$, $3p_1$, ..., $(\frac{n}{p_1})p_1$; their number is $\frac{n}{p_1}$. So the number of them which are not divisible by p_k is equal to

$$\phi_1(n) = n - \frac{n}{p_1} = n(1 - \frac{1}{p_1}). \tag{11.1}$$

Let us designate correspondingly with $\phi_\rho(n)$ the number of integers $\leq$ n which are not divisible by p_1, p_2, ..., p_ρ, (where $\rho \leq k$); the sought for number $\phi(n)$ is then $\phi_k(n)$.

Let us consider the integers $\leq$n which are not divisible by p_1, ..., p_ρ for some ρ, $1 \leq \rho \leq k-1$, and their number which we have designated by $\phi_\rho(n)$. These integers are of two types. The first type comprises those integers $\leq$ n which are not divisible by p_1, ..., p_ρ, $p_{\rho+1}$ and the number of this type is $\phi_{\rho+1}(n)$. The other type comprises those integers $\leq$ n which are not divisible by p_1, ..., p_ρ but are divisible by $p_{\rho+1}$. The integers of the second type are

1.11.2

found among the integers

$$1 \cdot p_{\rho+1}, \; 2p_{\rho+1} \;, \; \ldots, \; \frac{n}{p_{\rho+1}} \, p_{\rho+1}$$

of the form $\lambda p_{\rho+1}$ with $1 \leqslant \lambda \leqslant \frac{n}{p_{\rho+1}}$.   For an integer $\lambda p_{\rho+1}$ to be not divisible by any of $p_1, \ldots, p_\rho$, it is necessary and sufficient that λ not be divisible by any of the $p_1, \ldots, p_\rho$. The number of elements of the set $\{1, 2, \ldots, \frac{n}{p_{\rho+1}}\}$ which are not divisible by any of $p_1, \ldots, p_\rho$ is equal to

$\phi_\rho \left(\frac{n}{p_{\rho+1}}\right)$, using an obvious extension of our symbol. Thus we have found

$$\phi_\rho(n) = \phi_{\rho+1}(n) + \phi_\rho \left(\frac{n}{p_{\rho+1}}\right) , \text{ for } \rho = 1, 2, \ldots, k-1,$$

and thus

$$\phi_{\rho+1}(n) = \phi_\rho(n) - \phi_\rho \left(\frac{n}{p_{\rho+1}}\right) \text{ for } \rho = 1, 2, \ldots, k-1. \tag{11.2}$$

Now we apply this recursive formula repeatedly for $\rho = 1, 2, \ldots, k-1$ keeping (11.1) in mind, and we find

$$\phi_2(n) = \phi_1(n) - \phi_1\left(\frac{n}{p_2}\right) = n\left(1- \frac{1}{p_1}\right) - \frac{n}{p_2} \left(1- \frac{1}{p_1}\right)$$

$$= \left(n- \frac{n}{p_2}\right)\left(1- \frac{1}{p_1}\right) = n\left(1- \frac{1}{p_1}\right)\left(1- \frac{1}{p_2}\right)$$

$$\phi_3(n) = \phi_2(n) - \phi_2\left(\frac{n}{p_3}\right) = n\left(1- \frac{1}{p_1}\right)\left(1- \frac{1}{p_2}\right) - \frac{n}{p_3}\left(1- \frac{1}{p_1}\right)\left(1- \frac{1}{p_2}\right)$$

$$= \left(n- \frac{n}{p_3}\right)\left(1- \frac{1}{p_1}\right)\left(1- \frac{1}{p_2}\right) = n\left(1- \frac{1}{p_1}\right)\left(1- \frac{1}{p_2}\right)\left(1- \frac{1}{p_3}\right)$$

$$\cdot \quad \cdot \quad \cdot \quad \cdot$$

By mathematical induction with respect to k, for any natural number $n = p_1^{\alpha_1}, \ldots, p_k^{\alpha_k}$, we finally arrive at the formula

$$\phi(n) = \phi_k(n) = n\left(1-\frac{1}{p_1}\right)\left(1-\frac{1}{p_2}\right) \cdots \left(1-\frac{1}{p_k}\right).$$

Another expression of this formula is

$$\phi(n) = \phi(p_1^{\alpha_1} \cdots p_k^{\alpha_k}) = p_1^{\alpha_1-1}(p_1-1) \cdots p_k^{\alpha_k-1}(p_k-1)$$

<u>THEOREM</u> 1 If a and b are relatively prime natural numbers, then

$$\phi(ab) = \phi(a)\phi(b).$$

<u>Proof</u> Let

$$a = p_1^{\alpha_1} \cdots p_k^{\alpha_k} \text{ and } b = q_1^{\beta_1} \cdots q_\lambda^{\beta_\lambda}$$

where $\alpha_j > 1$ $(j=1, \ldots, k)$ and $\beta_i > 1$ $(i=1, \ldots, \lambda)$.

By the hypothesis that a and b are relatively prime, all the primes p_j are different from the primes q_i. Consequently the decomposition of ab into primes is

$$ab = p_1^{\alpha_1} \cdots p_k^{\alpha_k}q_1^{\beta_1} \cdots q_\ell^{\alpha_\ell}$$

and thus

$$\phi(ab) = ab\left(1-\frac{1}{p_1}\right) \cdots \left(1-\frac{1}{p_k}\right)\left(1-\frac{1}{q_1}\right) \cdots \left(1-\frac{1}{q_\ell}\right)$$

$$= a\left(1-\frac{1}{p_1}\right) \cdots \left(1-\frac{1}{p_k}\right)b\left(1-\frac{1}{q_1}\right) \cdots \left(1-\frac{1}{q_\ell}\right)$$

$$= \phi(a)\phi(b).$$

<u>THEOREM</u> 2 Let d be a positive divisor of n. Then the number of integers of the set $\{1, 2, \ldots, n\}$ which have greatest common divisor d with n is equal to $\phi\left(\frac{n}{d}\right)$.

<u>Proof</u> Let d be a divisor of n. The integers of the set $\{1, 2, \ldots, n\}$ which are multiples of d are

$$d, 2d, \ldots, \frac{n}{d}d; \text{ that is the } \lambda d, 1 < \lambda < \frac{n}{d}.$$

1.11.4

Since $(a\gamma, b\gamma) = (a,b)\gamma$ for $\gamma > 0$, and since

$$(kd,n) = (kd, \frac{n}{d} d) = (k, \frac{n}{d})d$$

the condition $d = (kd,n)$ is equivalent to $(k, \frac{n}{d}) = 1$, that is, the integer k must be relatively prime to $\frac{n}{d}$. So the number of multiples of d which have greatest common divisor d with n is equal to the number of k which are relatively prime to n, that is, $\phi(\frac{n}{d})$.

For example, for $n = 18$ and $d = 6$ we have $\phi(\frac{18}{6}) = \phi(3) = 2$. And indeed, among the integers 1, 2, ..., 17, 18 only the integers 6 and 12 have greatest common divisor 6 with 18. For $n = 18$ and $d = 2$ the number of those integers among 1, 2, ..., 18 which have greatest common divisor 2 with 18 is $\phi(\frac{18}{2}) = \phi(9) = 9(1-\frac{1}{3}) = 6$; and they are 2, 4, 8, 10, 14, 16.

THEOREM 3 For every natural number n we have

$$n = \sum_{d|n} \phi(d).$$

Proof Let $d_1 = 1, d_2, d_3, \ldots, d_k = n$ be the divisors of n in ascending order. The set of integers $\{1, 2, \ldots, n\}$ may be classified into subsets in the following way:

1st subset: integers which have GCD 1 with n

2nd subset: integers which have GCD d_2 with n

. . . . . .

j^{th} subset: integers which have GCD d_j with n

. . . . . .

k^{th} subset: integers which have GCD $d_k = n$ with n.

The corresponding number of elements in each subset is

$$\phi(n), \phi(\frac{n}{d_2}), \ldots, \phi(\frac{n}{d_j}), \ldots, \phi(\frac{n}{n}).$$

Consequently, since each integer of the set $\{1, 2, \ldots, n\}$ is in exactly one subset,

$$n = \phi(n) + \phi\left(\frac{n}{d_2}\right) + \ldots + \phi\left(\frac{n}{d_j}\right) + \ldots + \phi\left(\frac{n}{n}\right) = \sum_{j=1}^{k} \phi\left(\frac{n}{d_j}\right)$$

However, as the d_j run through the set of divisors of n, the $\frac{n}{d_j}$ run through the same set of divisors in the opposite order.   So

$$n = \sum_{d \mid n} \phi(d) \ .$$

12. DISTRIBUTION OF THE PRIME NUMBERS IN THE SEQUENCE OF NATURAL NUMBERS

We enumerate the prime numbers in increasing order

$$2 = p_1, \; 3 = p_2, \; 5 = p_3, \; 7 = p_4, \; \ldots, \; p_k < p_{k+1} , \ldots .$$

As we know, the sequence $p_1, p_2, p_3, \ldots, p_k, \ldots$ is nonterminating. In addition, we have the following:

THEOREM 1   The series $\displaystyle\sum_{n=1}^{\infty} \frac{1}{p_n}$ is divergent: $\displaystyle\lim_{k \to \infty} \sum_{j=1}^{k} \frac{1}{p_j} = +\infty.$

Proof   We regard as known from the elements of Mathematical Analysis that the series

$$\sum_{n=1}^{\infty} \frac{1}{n} \qquad \text{diverges,}$$

$$\sum_{m=1}^{\infty} \frac{1}{m^2} \qquad \text{converges}$$

and that for $|x| < 1$

$$\ln(1+x) = \frac{x}{1} - \frac{x^2}{2} + \ldots + (-1)^n \, \frac{x^n}{n} + \ldots .$$

Let n be an arbitrary natural number greater than 1, and

$$p_1, \; p_2, \; \ldots, \; p_k \le n < p_{k+1}$$

where, naturally, k depends on n. Each natural number a ≤ n has no prime divisors other than $p_1, \ldots, p_k$ and thus may be written in the form

$$a = p_1^{\alpha_1} \ldots p_k^{\alpha_k} \quad \text{with } \alpha_j \ge 0 \quad (j=1, \ldots, k). \tag{12.1}$$

We determine an integer r for which $2^r > n$; then $p_j^r > n$ for $j=1, \ldots, k$. Thus in the representation (12.1) of the numbers $1, \ldots, n$ we have $0 \le \alpha_j < r$ $(j=1, \ldots, k)$. We now note that for $j=1, \ldots, k$

$$\frac{1}{1 - \dfrac{1}{p_j}} > 1 + \frac{1}{p_j} + \frac{1}{p_j^2} + \ldots + \frac{1}{p_j^r}$$

since

$$\frac{1}{1-\frac{1}{p_j}} = \sum_{i=0}^{\infty} \frac{1}{p_j^i} = 1 + \frac{1}{p_j} + \frac{1}{p_j^2} + \ldots$$

Consequently, since $p_1^r \ldots p_k^r > n$, we will have

$$\prod_{j=1}^{k} \frac{1}{1-\frac{1}{p_j}} > \prod_{j=1}^{k} \left(1 + \frac{1}{p_j} + \frac{1}{p_j^2} + \ldots + \frac{1}{p_j^r}\right)$$

$$= \sum_{0 < \alpha < r} \frac{1}{p_1^{\alpha_1} \ldots p_k^{\alpha_k}}$$

$$> \sum_{m=1}^{n} \frac{1}{m} \qquad p_1,\ldots,p_k < n < p_{k+1} \qquad\qquad (12.2)$$

However, we have

$$\ln \frac{1}{1-\frac{1}{p_j}} = -\ln\left(1-\frac{1}{p_j}\right)$$

$$= -\left(-\frac{1}{p_j} - \frac{1}{2}\frac{1}{p_j^2} - \frac{1}{3}\frac{1}{p_j^3} - \ldots\right)$$

$$= \frac{1}{p_j} + \frac{1}{2}\frac{1}{p_j^2} + \frac{1}{3}\frac{1}{p_j^3} + \ldots$$

$$< \frac{1}{p_j} + \frac{1}{2p_j^2}\left(1 + \frac{1}{p_j} + \frac{1}{p_j^2} + \ldots\right)$$

$$= \frac{1}{p_j} + \frac{1}{2p_j^2}\left(\frac{1}{1-\frac{1}{p_j}}\right)$$

$$< \frac{1}{p_j} + \frac{1}{2p_j^2} \left(\frac{1}{1-\frac{1}{2}} \right)$$

$$= \frac{1}{p_j} + \frac{1}{p_j^2} \cdot$$

Thus we have found that

$$\ln \frac{1}{1-\frac{1}{p_j}} < \frac{1}{p_j} + \frac{1}{p_j^2} \qquad \text{for } j=1, \ldots, k$$

and thus

$$\sum_{j=1}^{k} \ln \frac{1}{1-\frac{1}{p_j}} < \sum_{j=1}^{k} \frac{1}{p_j} + \sum_{j=1}^{k} \frac{1}{p_j^2} \cdot$$

Hence

$$\sum_{j=1}^{k} \frac{1}{p_j} + \sum_{m=1}^{\infty} \frac{1}{m^2} > \sum_{j=1}^{k} \ln \frac{1}{1-\frac{1}{p_j}} = \ln \prod_{j=1}^{k} \frac{1}{1-\frac{1}{p_j}}$$

so, by (12.2),

$$\sum_{j=1}^{k} \frac{1}{p_j} + \sum_{m=1}^{\infty} \frac{1}{m^2} > \ln \left(\sum_{m=1}^{n} \frac{1}{m} \right) \cdot$$

Consequently

$$\sum_{j=1}^{k} \frac{1}{p_j} > \ln \left(\sum_{m=1}^{n} \frac{1}{m} \right) - \sum_{m=1}^{\infty} \frac{1}{m^2} \cdot$$

But as $n \to +\infty$, we also have $k \to +\infty$ and $\sum_{m=1}^{n} \frac{1}{m} \to +\infty$.

Hence

$$\lim_{n \to \infty} \ln \left(\sum_{m=1}^{n} \frac{1}{m} \right) = +\infty$$

and it follows that

$$\lim_{n \to \infty} \sum_{j=1}^{n} \frac{1}{p_j} = +\infty$$

which was to be proved.

For non-negative real x we will set

$$\pi(x) = (\text{number of prime numbers} \leq x).$$

So we have

$$\pi(x) = 0 \quad \text{for} \quad 0 \leq x < 2$$

$$\pi(x) = 1 \quad \text{for} \quad 2 \leq x < 3$$

$$\pi(x) = 2 \quad \text{for} \quad 3 \leq x < 5$$

$$\pi(x) = 3 \quad \text{for} \quad 5 \leq x < 7$$

$$\pi(x) = 4 \quad \text{for} \quad 7 \leq x < 11 \quad \text{etc.}$$

Obviously $\pi(x)$ is an increasing discontinuous function in the domain $0 \leq x < +\infty$ with points of discontinuity at the prime numbers 2, 3, 5, ... If $p_k \leq x < p_{k+1}$ then $\pi(x) = k$.

Without proof we will present a number of propositions regarding the function $\pi(x)$.

<u>THEOREM</u> 2 We have the following limiting relationships

$$\lim_{x \to \infty} \frac{\pi(x)}{x/\ln x} = \lim_{x \to \infty} \frac{\pi(x) \ln x}{x} = 1. \tag{12.3}$$

We have, for example, for various specific values of x

$$\pi(500) = 95. \qquad \frac{\pi(500) \ln 500}{500} \approx \frac{95 \cdot 6.21461}{500} \approx 1.181$$

$$\pi(1000) = 168. \qquad \frac{\pi(1000) \ln 1000}{1000} \approx \frac{168 \cdot 6.90776}{1000} \approx 1.161$$

$$\pi(2000) = 303. \qquad \frac{\pi(2000) \ln 2000}{2000} \approx \frac{303 \cdot 7.60091}{2000} \approx 1.151$$

$$\pi(5000) = 669. \qquad \frac{\pi(5000) \ln 5000}{5000} \approx \frac{669 \cdot 8.51720}{5000} \approx 1.140.$$

From Theorem 2 we conclude that

$$\pi(x) = \frac{x}{\ln x}(1+\varepsilon_x) \qquad\qquad \lim_{x\to\infty}\varepsilon_x = 0$$

__THEOREM__ 3 If we set $\sum_{p\leqslant x}\ln p = \theta(x)$, then $\lim_{x\to\infty}\dfrac{\theta(x)}{x} = 1$

Consequently

$$\theta(x) = x(1 + \varepsilon'_x) \quad\text{where}\quad \lim_{x\to\infty}\varepsilon'_x = 0.$$

__THEOREM__ 4 If we set $\sum_{p\leqslant x}\dfrac{1}{p} = \psi(x)$, then

$$\psi(x) = \sum_{p\leqslant x}\frac{1}{p} = \ln\ln x + B + \tilde{\varepsilon}_x \quad\text{with}\quad \lim_{x\to\infty}\tilde{\varepsilon}_x = 0,$$

where B is a constant.

For comparison we mention (still without proof) that

$$\sum_{n\leqslant x}\frac{1}{n} = \ln x + C + \varepsilon''_x$$

where C is Euler's constant (C = 0.5772157...) and $\lim_{x\to\infty}\varepsilon''_x = 0$.

We have $\pi(p_n) = n$ and $\lim_{n\to\infty}p_n = +\infty$, so according to Theorem 2,

$$\frac{\pi(p_n)\ln p_n}{p_n} = \frac{n\ln p_n}{p_n} \to 1 \qquad\text{for } n\to +\infty. \tag{12.4}$$

Thus

$$\ln\frac{\pi(p_n)\ln p_n}{p_n} = \ln n + \ln\ln p_n - \ln p_n \to \ln 1 = 0 \quad\text{for } n\to +\infty$$

and it follows that

$$\frac{\ln n}{\ln p_n} + \frac{\ln\ln p_n}{\ln p_n} - 1 \to 0 \qquad\text{for } n\to +\infty .$$

Consequently, since $\lim\limits_{n\to\infty} \dfrac{\ln \ln p_n}{\ln p_n} = 0$,

$$\lim_{n\to\infty} \frac{\ln n}{\ln p_n} = 1.$$

(12.5)

Finally, making use of (12.4) and (12.5) we find

$$\lim_{n\to\infty} \frac{n \ln n}{p_n} = \lim_{n\to\infty} \frac{n \ln p_n}{p_n} \cdot \frac{\ln n}{\ln p_n} = 1 \cdot 1 = 1 .$$

<u>Theorem</u> 5* For any natural number a > 4, between a and 2a-2 there exists at least one prime p: a < p < 2a-2.

<u>Theorem</u> 6† If $(d,r) = 1$ then the arithmetic progression

$$r, \ r + d, \ r + 2d, \ \ldots, \ r + md, \ r + (m+1)d, \ \ldots$$

contains infinitely many prime numbers.

For example, there are infinitely many primes of the form 4n+1 and infinitely many of the form 4n+3, where n is a natural number.

There exist integer polynomials f(n) which give prime numbers f(n) for whole series of values of n; for example:

the number $f_1(n) = n^2 + n + 17 = n(n+1) + 17$

is prime for n = 0, 1, 2, ..., 15;

the number $f_2(n) = n(n+1) + 41$

is prime for n = 0, 1, 2, ..., 39;

the number $f_3(n) = 2n^2 + 29$

is prime for n = 0, 1, 2, ..., 28.

---

* For historical reasons, this is called Bertrand's postulate.
† Dirichlet's theorem on primes in an arithmetic progression.

There does not exist, however, an integer polynomial $f(n)$ of the natural number n such that $f(n) = p_n$ for $n = 1, 2, \ldots$. This follows easily from the limiting relationship (12.4).

To conclude this section we present three questions to which, at this time, no answer is known.

I. Two primes p, q with difference $p - q = 2$ are called twin primes. For example, $\{3,5\}$, $\{11,13\}$, $\{17,19\}$, $\ldots$, $\{149,151\}$, $\ldots$ are twin primes.

The unanswered question is: Is the number of twin primes finite or infinite?

II. There exist even integers which are the sum of two primes, for example:

$$4 = 2+2, \ 6 = 3+3, \ 8 = 3+5, \ 10 = 5+5, \ 12 = 5+7, \ 14 = 3+11, \ 16 = 3+13,$$
$$18 = 5+13 = 7+11, \ 20 = 7+13, \ 22 = 11+11 = 17+5, \ 24 = 5+19.$$

The unanswered question is: Can every even integer greater than 2 be decomposed into a sum of two primes?

III. As was proved by Gauss, the regular polygon with 2^m+1 sides is constructible with compass and straightedge when 2^m+1 is a prime number. For example, the 17-gon is constructible since $17 = 2^4+1 = 2^{2^2}+1$.

But for 2^m+1 to be a prime, it is necessary that m have no odd prime factors. Indeed, if $m = pq$ with p an odd prime, then

$$2^m+1 = 2^{pq}+1 = (2^q)^p+1$$
$$= (2^q+1)[(2^q)^{p-1} - (2^q)^{p-2}+ \ldots -(2^q)^1+1]$$

and thus 2^m+1 cannot be prime. Therefore, for 2^m+1 to be prime, m must be of the form 2^n with n a non-negative integer. We have, for example

$$2^{2^0}+1 = 3, \quad 2^{2^1}+1 = 5, \quad 2^{2^2}+1 = 17, \quad 2^{2^3}+1 = 257, \quad 2^{2^4}+1 = 65,537$$

which are all primes, but

$$2^{2^5}+1 = 4{,}294{,}967{,}297 = 641 \cdot 6{,}700{,}417.$$

Hence the condition $m = 2^n$ is not sufficient for 2^m+1 to be a prime.

The unanswered question is: Is the number of primes of the form $2^{2^n}+1$ finite or infinite?

1.P.1

PROBLEMS FOR CHAPTER I

1. Show that if $c > 0$ then $(ac,bc) = (a,b)c$.

2. Show that if p is prime and $p \nmid r$ then $(r,p^e m) = (r,m)$.

3. Show that the greatest common divisor of a, b, and c is $((a,b),c)$.

4. Show by example that, in general,

    $[\text{LCM of } a,b,c] \cdot [\text{GCD of } a,b,c] \neq abc$.

    Under what conditions will we have such an equality?

5. Show that if $c \mid ab$ and $(c,a) = d$ then $c \mid db$.

6. Show that the least common multiple of a, b, and c is the product abc

    divided by the greatest common divisor of bc, ac, and ab.

7. Prove that $1 + \frac{1}{2} + \frac{1}{3} + \ldots + \frac{1}{n}$ is not an integer $n > 1$.

8. Define $\operatorname{ord}_p a = n \iff p^n \mid a$ but $p^{n+1} \nmid a$. Show that

    a) if $\operatorname{sgn}(a) = \operatorname{sgn}(b)$ and $\operatorname{ord}_p(a) = \operatorname{ord}_p(b)$ for all primes p then $a = b$.

    b) $\operatorname{ord}_p ab = \operatorname{ord}_p a + \operatorname{ord}_p b$

    c) $\operatorname{ord}_p(a+b) \geqslant \min\{\operatorname{ord}_p a, \operatorname{ord}_p b\}$

    d) if $\operatorname{ord}_p a \neq \operatorname{ord}_p b$ then $\operatorname{ord}_p(a+b) = \min\{\operatorname{ord}_p a, \operatorname{ord}_p b\}$

    e) $\operatorname{ord}_p((a,b)) = \min(\operatorname{ord}_p a, \operatorname{ord}_p b)$; $\operatorname{ord}_p(\operatorname{LCM}(a,b)) = \max\{\operatorname{ord}_p a, \operatorname{ord}_p b\}$

    f) Show, using ord, that $(a+b, \operatorname{LCM}(a,b)) = (a,b)$.

9. Define ord_p for fractions r/s; $r,s \in \mathbb{Z}$. Do 8abcd still hold?

10. Consider the subset of the rational numbers r/s for which $\operatorname{ord}_p r/s \geqslant 0$.

    Show this system is closed under addition and multiplication.

    Characterize those elements α of the system for which there exists a β

    such that $\alpha\beta = 1$. (Such elements are analogous to ± 1 in $\mathbb{Z}$). Are there

    any primes in this system? Do elements have unique factorization?

The following problems develop a notation of Gauss and Dirichlet useful in the theory of continued fractions and elsewhere.

11. Define $[a_1, a_2, \ldots, a_n]$ by

$$[a_1] = a_1$$
$$[a_1, a_2] = a_1 a_2 + 1$$
$$[a_1, \ldots, a_n] = [a_1, \ldots, a_{n-2}] + [a_1, \ldots, a_{n-1}] a_n$$

Show by induction that

a) $[a_1, \ldots, a_n] = a_1 [a_2, \ldots, a_n] + [a_3, \ldots, a_n]$ for $n \geqslant 3$

b) $[a_1, \ldots, a_n] = [a_n, \ldots, a_1]$

c) $[-a_1, -a_2, \ldots, -a_n] = (-1)^n [a_1, \ldots, a_n]$.

12. Suppose that $\quad r_3 = a_1 r_2 + r_1$

$$r_4 = a_2 r_3 + r_2$$

$$\cdots\cdots\cdots\cdots\cdots\cdots$$

$$r_{n+2} = a_n r_{n+1} + r_n \ .$$

Show by induction that,

a) $r_{n+2} = [a_1, \ldots, a_n] r_2 + [a_2, \ldots, a_n] r_1$ for $n \geqslant 2$

and hence, by rearrangement of the equations,

b) $r_1 = [-a_n, -a_{n-1}, \ldots, -a_1] r_{n+2} + [-a_{n-1}, \ldots, -a_1] r_{n+1}$

$$= (-1)^n ([a_1, \ldots, a_n] r_{n+1} - [a_1, \ldots, a_{n-1}] r_{n+2}).$$

13. In 12) let $r_1 = 0$, $r_2 = 1$. What are r_j, $j = 3, \ldots, n+2$? By omitting the first equation, conclude that, for $n \geqslant 3$

$$(-1)^n = [a_1, \ldots, a_n][a_2, \ldots, a_{n-1}] - [a_1, \ldots, a_{n-1}][a_2, \ldots, a_n].$$

14. Recall that the Euclidean algorithm for $d = (a_1, a_2)$, $a_1, a_2 > 0$, has the form

$$a_i = q_i a_{i+1} + a_{i+2} \qquad 0 < a_{i+2} < a_{i+1} \qquad i = 1,\ldots,n$$

$$a_{n+1} = q_{n+1} a_{n+2}$$

where $d = a_{n+2}$.

a) Show $d = (-1)^n([q_1,\ldots,q_n]a_2 - [q_2,\ldots,q_n]a_1)$

$$a_1 = d[q_1,\ldots,q_{n+1}], \quad a_2 = d[q_2,\ldots,q_{n+1}] \ .$$

This gives us an efficient computational method of finding r and s such

that $ra_1 + sa_2 = (a_1,a_2)$.

Remark. An efficient means to organize the computation of $[q_1,q_2,\ldots,q_j]$ and

$[q_2,\ldots,q_j]$ for $j = 1,2,\ldots$ is by means of the following tabular form

		q_1	q_2	q_3	q_4
0	1	q_1	$[q_1,q_2]$	$[q_1 q_2 q_3]$	
1	0	1	$[q_2]$	$[q_2 q_3]$	

where the entry for q_4, for example, is found by multiplying q_4 times the

entry immediately to the left and adding the entry one further position

to the left. This is done independently for each row. For example,

using the quotients 2, 1, 4, 3, 2 found when using the Euclidean

Algorithm to compute (41,18), one finds

	2	3	1	1	2
0	1	2	7	9	16	41
1	0	1	3	4	7	18

and the greatest common divisor is

$$d = (-1)^4([2\ 3\ 1\ 1]18 - [3\ 1\ 1]41) = 16 \cdot 18 - 7 \cdot 41 = 1.$$

Note the final column provides a check.

If no check is desired and the intermediate $[q_1,\ldots,q_j]$ are of no interest, it is more efficient to compute $[q_n,\ldots,q_1]$ and $[q_n,\ldots,q_2]$ with one row:

$$
\begin{array}{cccc}
1 & 1 & 3 & 2 \\
\end{array}
$$

$$
\overline{0 \quad 1 \ \big|\ 1 \ \big|\ 2 \ \big|\ 7 \ \big|\ 16} \ .
$$

b) Use this method to express the GCD of 175 and 63 as a linear combination of 175 and 63.

15. Show by induction that

$$[a_1,\ldots,a_n] = [a_1,\ldots,a_{k-2}][a_{k+1},\ldots,a_n]+[a_1,\ldots,a_{k-1}][a_k,\ldots,a_n] \ .$$

(Induction is on $n > k+1$, k fixed). This may be viewed as a generalization of problem 11.

16. The proof of Theorem 1 of §12 uses the fact that there are infinitely many primes in a very inessential way. Restructure the proof slightly so that it becomes a proof that there are infinitely many primes.

17. Prove that, for $s > 1$, $\displaystyle\sum_{n=1}^{\infty} \frac{1}{n^s} = \Pi(1 - \frac{1}{p^s})^{-1}$, where the product is the infinite product over all primes.

Hint: Consider equation (12.2), where m and p_j are replaced by m^s and p_j^s. Equation (12.2) will then give a lower bound for the product.

It is necessary to have also the upper bound $\displaystyle\sum_{m=1}^{\infty} \frac{1}{m^s}$.

18. Use the integral test to show that $\displaystyle\sum_{m=1}^{n} \frac{1}{m} > \ln n + \frac{1}{n+1}$ and $\displaystyle\sum \frac{1}{m^2} < 2$. Show that $\displaystyle\sum_{j=1}^{k} \frac{1}{p^j} > \ln \ln n - 2$ where $p_k \leqslant n < p_{k+1}$.

19. Prove that $\phi(p^e) = p^e - p^{e-1}$ by demonstrating that there are exactly p^{e-1} multiples of p between 1 and p^e.

The following problems develop the Möbius μ-Function and apply it to Euler's ϕ-Function. All integers n and d are assumed positive.

20. Let $n = 1$ or $n = p_1^{\alpha_1} p_2^{\alpha_2} \ldots p_k^{\alpha_k}$ where $1 \leqslant \alpha_j$, $j = 1,\ldots,k$.

Define $\mu(n)$ for $n \in \mathbb{N}$

$$\mu(n) = \begin{cases} 0 & \text{If any } \alpha_j \geqslant 2, \ j = 1,\ldots,k \\ 1 & \text{If } n = 1 \\ (-1)^k & \text{If } n > 1 \text{ and all } \alpha_j = 1, \ j = 1,\ldots,k \end{cases}$$

Show $\displaystyle\sum_{d \mid n} \mu(d) = \begin{cases} 1 & \text{for } n = 1 \\ 0 & \text{for } n > 1 \end{cases}$

Hint: Consider d in the form $p_1^{\beta_1} \ldots p_k^{\beta_k}$, $\beta_j = 0$ or 1, count the possibilities using binomial coefficients, and show the sum is a power of $(1-1)$ when $n > 1$.

21. For two functions f, g defined on $\mathbb{N}$ with values in the complex numbers, define the convolution

$$f{*}g(n) = \sum_{d \mid n} f\left(\tfrac{n}{d}\right) g(d).$$

Prove convolution is an associative operation: $(f{*}g){*}h = f{*}(g{*}h)$ and a commutative operation $f{*}g = g{*}f$.

<u>Hint</u> Rewrite the definition using two divisors and generalize.

22. a) Let $\delta(n) = \begin{cases} 1 & \text{for } n = 1 \\ 0 & \text{for } n > 1 \end{cases}$. Show $f{*}\delta = f$.

b) Let $I(n) = 1$ for all n. Show $I{*}\mu = \delta$.

Show $I{*}I = $ (Number of natural divisors of n).

c) Let $J(n) = n$. Show $I{*}J = Z$, where $Z(n)$ is the sum of the divisors of n. (See Thm 3 of §5).

23. Let $F(n) = \displaystyle\sum_{d \mid n} f(d)$. Show $F = I{*}f$. Use 22 and the associative law to prove the MÖBIUS INVERSION FORMULA

$$f(n) = \sum_{d \mid n} \mu\left(\tfrac{n}{d}\right) F(d) .$$

24. Prove $n = \sum_{d|n} \phi(d)$ by reducing the fractions $\frac{j}{n}$, $j = 1,\ldots,n$.

Use the Möbius inversion formula to get

$$\phi(n) = n \sum_{d|n} \frac{\mu(d)}{d} = n \left(1 - \sum_{p_i|d} \frac{1}{p_i} + \sum_{p_i,p_j|d} \frac{1}{p_i p_j} - \cdots\right)$$

where the sums are taken over products of distinct primes. Show this

gives the formula for Euler's ϕ-Function.

25. Let $f(n) = \sum_{\substack{(a,n)=1 \\ 1 < a < n}} a$, $h(n) = \frac{n+1}{2}$. Show by considering the reduced

fractions $\frac{j}{n}$, $1 < j < n$, that $\sum_{d|n} \frac{f(d)}{d} = h(n)$.

Use the Möbius inversion formula to show

$$f(n) = \frac{1}{2} n\phi(n).$$

Chapter 2
Congruences

13. THE CONCEPT OF CONGRUENCE AND BASIC PROPERTIES

An integer a is called _congruent_ to an integer b modulo a natural number m if and only if the difference a-b is divisible by m: a-b=mq with integer q. We symbolize the above relationship by

$$a \equiv b \pmod{m}.$$

This relationship of congruence is an equivalence relation, because the three defining conditions of an equivalence relation hold:

I) $a \equiv a \pmod{m}$ (reflexivity)
II) $a \equiv b \pmod{m} \Rightarrow b \equiv a \pmod{m}$ (symmetry)
III) $[a \equiv b \pmod{m}$ and $b \equiv c \pmod{m}] \Rightarrow a \equiv c \pmod{m}$ (transitivity).

In accordance with the above observations the set $\mathbb{Z}$ of integers is divided into equivalence classes modulo m. The number of these classes is equal to m. In fact, the division of an integer a by m leaves as remainder one of the numbers $0, 1, \ldots, m-1$, and two integers are congruent mod m if and only if they leave the same remainder when divided by m. Hence there are as many equivalence classes as there are remainders. An integer a is thus congruent to one and only one of the m numbers $0, 1, \ldots, m-1$.

We now formulate a number of theorems for congruences.

<u>THEOREM</u> 1 We may add or subtract two congruences with the same modulus m:

$$\left.\begin{array}{l} a_1 \equiv b_1 \pmod{m} \\[1ex] a_2 \equiv b_2 \pmod{m} \end{array}\right\} \Rightarrow a_1 \pm a_2 \equiv b_1 \pm b_2 \pmod{m}.$$

<u>Proof</u> Indeed,

$$\left.\begin{array}{l} a_1 - b_1 = q_1 m \\[1ex] a_2 - b_2 = q_2 m \end{array}\right\} \Rightarrow a_1 \pm a_2 - (b_1 \pm b_2) = (q_1 \pm q_2) m.$$

<u>THEOREM</u> 2 $a \equiv b \pmod{m} \Rightarrow ga \equiv gb \pmod{m}$ for $g \in \mathbb{Z}$

    <u>Proof</u> Indeed,

$$a \equiv b \pmod{m} \Rightarrow m \mid a-b \Rightarrow m \mid g(a-b) = ga - gb$$
$$\Rightarrow ga \equiv gb \pmod{m}$$

<u>THEOREM</u> 3   We may multiply corresponding sides of two congruences with the same modulus m

$$\left.\begin{array}{l} a_1 \equiv b_1 \pmod{m} \\ a_2 \equiv b_2 \pmod{m} \end{array}\right\} \Rightarrow a_1 \cdot a_2 \equiv b_1 \cdot b_2 \pmod{m},$$

    <u>Proof</u> According to theorem 2

$$a_1 \equiv b_1 \pmod{m} \Rightarrow a_1 \cdot a_2 \equiv b_1 \cdot a_2 \pmod{m}$$
$$a_2 \equiv b_2 \pmod{m} \Rightarrow b_1 \cdot a_2 \equiv b_1 \cdot b_2 \pmod{m}.$$

    Hence $a_1 \cdot a_2 \equiv b_1 \cdot b_2 \pmod{m}$ by transitivity.

The above theorems lead to the following general proposition

<u>THEOREM</u> 4   Let $F(x_1, x_2, \ldots, x_n)$ be a polynomial in the variables $x_1, x_2, \ldots, x_n$ with integer co-efficients; that is, a sum $F(x_1, x_2, \ldots, x_n) = \sum c\, x_1^{\rho_1} \ldots x_n^{\rho_n}$ of finitely many terms of the form $cx_1^{\rho_1} \ldots x_n^{\rho_n}$ where $\rho_1, \ldots, \rho_n$ are non-negative integers and c is an integer. Then the congruences

$$a_1 \equiv b_1, \ldots, a_n \equiv b_n \pmod{m}$$

imply

$$F(a_1, \ldots, a_n) \equiv F(b_1, \ldots, b_n) \pmod{m}.$$

14. CRITERIA OF DIVISIBILITY

I. Let $m = 2$ or 5. A natural number n may be written in the decimal system of notation in the form $z = x_0 + 10x_1$ where x_0 is the number of units, $0 \leqslant x_0 \leqslant 9$ and x_1 is the number of tens, x_1 any non-negative integer. Since 10 is divisible by $m = 2$ or 5, it follows that $z \equiv x_0 \pmod{2 \text{ or } 5}$. Thus an integer n in divisible by 2 or 5 if and only if the number x_0 of its units is divisible by 2 or 5; that is, $x_0 = 0,2,4,6,8$ for $m = 2$ and $x_0 = 0,5$ for $m = 5$.

II. Let $m = 3$ or 9.

We have $z = x_0 + 10\,x_1 + 10^2\,x_2 + \ldots + 10^n\,x_n$ where $x_0, x_1, \ldots, x_n$ are the digits of the representation of z in the decimal system. However

$$10 \equiv 1 \pmod{3 \text{ or } 9},$$

so

$$10^n \equiv 1^n \equiv 1 \pmod{3 \text{ or } 9}.$$

Therefore

$$z \equiv x_0 + x_1 + x_2 + \ldots + x_n \pmod{3 \text{ or } 9}.$$

From this follows the well known criterion of divisibility:

An integer z is divisible by 3 or 9 if and only if the sum of its digits (in the decimal expansion) is divisible by 3 or 9.

III. $m = 11$

We have

$$10 \equiv -1 \pmod{11} \Rightarrow 10^n \equiv (-1)^n \pmod{11}.$$

so for $z = x_0 + 10\, x^1 + 10^k\, x_k$ we have

$$z \equiv x_0 - x_1 + x_2 - \ldots + (-1)^k\, x_k \pmod{11}.$$

Consequently z is divisible by 11 if and only if the alternating sum of is digits is divisible by 11. For example, 7321 is not divisible by 11 because $1 - 2 + 3 - 7 = -5$ is not divisible by 11.

IV. $m = 7$

Let $z = x_0 + 10\, x_1$, $0 \leqslant x_0 \leqslant 9$ and x_1 a non negative integer. Then z is divisible by 7 if and only if $2z = 2x_0 + 20x_1$ is divisible by 7. However $20 \equiv -1 \pmod 7$ so

$$2z \equiv 2x_0 - x_1 \equiv -(x_1 - 2x_0) \pmod 7.$$

Consequently z is divisible by 7 if and only if $x_1 - 2x_0$ is divisible by 7. For example 854 is divisible by 7 because $85 - 2 \cdot 4 = 77$ is divisible by 7.

V. PROBLEM: On what day of the week does a given date fall?

Solution We note first that those years with 366 days (according to the Gregorian Calendrical system) are years whose numbers n are divisible by 4 except for those divisible by 100; however, if the year is divisible by 400, it again has 366 days. For example, 1900 had 365 days, but 2000 will have 366.

Secondly, we note that the number of natural numbers $\leqslant n$ which are divisible by a natural number d is equal to the largest integer u such that $ud \leqslant n$; that is, $u \leqslant \frac{n}{d}$. We denote this u by $[\frac{n}{d}]$.

Now let m be the number of a day in the year with number n + 1. If we number the days since 1 Jan of year 1 (a Monday), the given day will be the day number

$$H = n\cdot 365 + \text{(number of previous years with 366 days)} + m$$

$$= n\cdot 365 + \left[\frac{n}{4}\right] - \left[\frac{n}{100}\right] + \left[\frac{n}{400}\right] + m.$$

Coding the days of the week by

0 = Sunday,   1 = Monday, . . ., 5 = Friday,   6 = Saturday,

it is clear that the day of the week will be given by the congruence class of H (mod 7). Calculations may be simplified by finding the residue class of n (mod 7) and noting that $365 \equiv 1$ (mod 7).

<u>Example</u>: 26 July 1980. Then n=1979 and m=208 (including 29 Feb 1980).
We have $1979 \equiv 5$ (mod 7). So

$$H \equiv 5\cdot 1 + \left[\frac{1979}{4}\right] - \left[\frac{1979}{100}\right] + \left[\frac{1979}{400}\right] + 208$$

$$\equiv 5 + 494 - 19 + 4 + 208 \equiv 692 \equiv 6 \quad \text{(mod 7)}$$

indicating that 26 July 1980 was a Saturday.

15. FURTHER THEOREMS ON CONGRUENCES

<u>THEOREM</u> 1 The congruence $ka \equiv kb \pmod{m}$ with k a natural number and $(k,m) = d$ is equivalent to $a \equiv b \pmod{\frac{m}{d}}$.

<u>Proof</u> We have $k = dk_1$, $m = dm_1$ with $(k_1, m_1) = 1$.

The congruence $ka \equiv kb \pmod{m}$ means $k(a-b) = ka-kb = mg$

with integer g. Thus $k_1 d (a-b) = m_1 dg$, so

$k_1 (a-b) = m_1 g$. From this it follows that $k_1 (a-b)$ is divisible by

m_1 and, since $(k_1, m_1) = 1$, that $a-b$ is divisible by m_1. Hence

$$a \equiv b \pmod{m_1 = \tfrac{m}{d}}.$$

On the other hand, if $a \equiv b \pmod{\frac{m}{d}}$, so that

$$a-b = \tfrac{m}{d} g_1 \text{ with integer } g_1$$

then

$$dk_1 a - dk_1 b = dk_1 (a-b) = dk_1 \tfrac{m}{d} g_1 = m k_1 g_1,$$

since $dk_1 = k$, and setting $g = k_1 g_1$ we have

$$ka - kb = mg$$

and

$$ka \equiv kb \pmod{m}.$$

<u>Corollary</u> If $ka \equiv kb \pmod{m}$ and $(k,m) = 1$ then $a \equiv b \pmod{m}$.

<u>Proof</u> This is a consequence of Theorem 1 with $d = 1$.

<u>Remark</u> From $ka \equiv kb \pmod{m}$ we cannot infer $a \equiv b \pmod{m}$.

For example $2 \cdot 6 \equiv 2 \cdot 4 \pmod{4}$

does not imply that $6 \equiv 4 \pmod{4}$, though it does imply

$6 \equiv 4 \pmod{\frac{4}{2} = 2}$.

THEOREM 2 $a \equiv b \pmod m$ and $d \mid m$ implies $a \equiv b \pmod d$.

The proof is immediate.

THEOREM 3 Let $a \equiv b \pmod{m_1}$ and $a \equiv b \pmod{m_2}$, and let m be

the least common multiple of m_1 and m_2.

Then $a \equiv b \pmod m$.

> Proof By hypothesis $a-b$ is a common multiple of m_1 and m_2, and
>
> hence a multiple of their least common multiple m.

THEOREM 4 From $a \equiv b \pmod m$ it follows that $(a, m) = (b, m)$.

> Proof The hypothesis means that $a = b + km$ with integer k.
>
> Each common divisor of a and m is a divisor of b and each common
>
> divisor of b and m is a divisor of a. So the greatest common
>
> divisors (a, m) and (b, m) are equal.

16. RESIDUE CLASSES mod m.

As we have seen, the set $\mathbb{Z}$ is divided into equivalence classes, (mod m) also called residue classes, each of which comprises those integers congruent to one another. Thus we have the following classes mod m

class congruent to 0: ...-2m, -m, 0, m, 2m...,

class congruent to 1: ...-2m + 1, -m + 1, 1, m + 1, 2m + 1,...,

class congruent to m-1: ... -2m + m - 1, -m + m - 1,0 + m - 1, m + m - 1,... .

COMPLETE SYSTEM OF RESIDUES mod m

Any set of m integers containing exactly one element from each residue class is called a complete system of residues mod m.

Thus any integer is congruent with one and only one element of a complete system of residues.

The members of a complete system of residues are incongruent, one to the other. On the other hand, m integers incongruent to one another form a complete system of residues mod m.

We distinguish the following complete systems of residues.

1. The least non-negative system of residues mod m: $\{0, 1, 2,..., m-1\}$.

2. The least positive system of residues mod m: $\{1, 2,..., m\}$.

3. The complete system of absolute least residues mod m:

For $m \equiv 1 \pmod 2$, that is, odd m
$$\{-\tfrac{m-1}{2},..., -1, 0, 1,..., \tfrac{m-1}{2}\}.$$

For $m \equiv 0 \pmod 2$, that is, even m
$$\{ -(\tfrac{m}{2} - 1),...,-1, 0, 1,...,\tfrac{m}{2} - 1, \tfrac{m}{2}\}$$
or $\{ -\tfrac{m}{2}, ..., -1, 0, 1,...,\tfrac{m}{2} - 1 \}.$

The numbers of a class mod m all have the same greatest common divisor with m, as we know from Theorem 4 of § 15. Specifically, $\phi(m)$ residue classes are formed out of integers whose greatest common divisor with m is 1. A residue class whose elements are relatively prime to m, that is, whose elements have greatest common divisor 1 with m, will be called <u>relatively prime to m</u>. Thus there are $\phi(m)$ relatively prime residue classes.

A set $\{ r_1, r_2, \ldots, r_{\phi(m)} \}$ of integers which contains exactly one element from each relatively prime residue class mod m is called a reduced system of residues mod m. For example, a reduced system of residues mod 5 is $\{1, 2, 3, 4\}$ and a reduced system of residues mod 12 is $\{1, 5, 7, 11\}$.

The reduced system of least positive residues mod m is made up of those integers $1, 2, \ldots, m$ which are relatively prime to m.

17. THE THEOREM OF FERMAT

__THEOREM__ Let p be a prime number and $(r,p) = 1$ (that is, $p \nmid r$). Then

$$r^{p-1} - 1 \equiv 0 \pmod p$$

__Proof__ We first note that the integers (they are integers because they represent the number of combinations of p things taken k at a time)

$$P_k = \binom{p}{k} = \frac{p\,(p-1)\ldots(p-k+1)}{1.2\ldots k} \quad \text{for } 1 < k < p-1$$

are divisible by p. Actually, the numerator of the fraction contains the prime factor p which, however, is not a divisor of the denominator since $1 < k < p-1$.

We now have

$$(a_1 + a_2)^p = a_1^{\,p} + p_1 a_1^{\,p-1} a_2 + \ldots + p_k\, a_1^{\,p-k} a_2^{\,k} + \ldots + p_{p-1}\, a_1\, a_2^{\,p-1} + a_2^{\,p}$$
$$\equiv a_1^p + a_2^p \pmod p.$$

Using mathematical induction this congruence generalizes to

$$(a_1 + a_2 + \ldots + a_r)^p \equiv a_1^p + a_2^p + \ldots + a_r^p \pmod p \quad \text{for } r \in \mathbb{N}.$$

Next we set $a_1 = a_2 = \ldots = a_r = 1$ and find

$$r^p \equiv 1^p + \ldots + 1^p$$

$$\equiv 1 + 1 + \ldots + = r \pmod p \quad \text{for } r \in \mathbb{N}. \tag{17.1}$$

For $r = 0$, $0^p \equiv 0 \pmod p$.

For $r = -n < 0$ where n is a natural number and p is an odd prime we have, in agreement with (17.1),

$$r^p = (-n)^p = -n^p \equiv -n = r \pmod p.$$

For $r = -n < 0$ where n is a natural number and $p = 2$ we have

$$r^p = r^2 = (-n)^2 = n^2 \equiv n \equiv -n = r \pmod 2. \tag{17.2}$$

Thus we have $r^p \equiv r \pmod p$ for any integer r. Specializing to the case $(r, p) = 1$, we may divide (17.1) by r (by the corollary of § 15) and thus we have

$$r^{p-1} - 1 \equiv 0 \pmod p. \tag{17.3}$$

18. GENERALIZED THEOREM OF FERMAT

<u>THEOREM</u> Let m be a natural number and r an integer with $(r, m) = 1$. Then

$$r^{\phi(m)} \equiv 1 \pmod{m}.$$

<u>Remark</u> The following proof constitutes, when m is prime, a second proof of the theorem of Fermat of § 17.

<u>Proof</u> We take a reduced system of residues mod m,

$\{r_1, \ldots, r_{\phi(m)}\}$. For each r with $(r, m) = 1$, the system

$$\{rr_1, rr_2, \ldots, rr_{\phi(m)}\}$$

is also a reduced system of residues mod m.

Indeed, there are $\phi(m)$ elements, each relatively prime to m (Theorem 1 of § 9) and they are incongruent since

$rr_i \equiv rr_j \pmod{m}$ and $(r, m) = 1$ implies $r_i \equiv r_j \pmod{m}$.

Thus each of rr_k $(1 \leqslant k \leqslant \phi(m))$ is congruent to exactly one

of $r_1, r_2, \ldots, r_{\phi(m)}$. Thus

$$(rr_1)(rr_2)\ldots(rr_{\phi(m)}) \equiv r_1 r_2 \ldots r_{\phi(m)} \pmod{m}, \qquad (18.1)$$

which may be written

$$r^{\phi(m)}(r_1 r_2 \ldots r_{\phi(m)}) \equiv r_1 r_2 \ldots r_{\phi(m)} \pmod{m}. \qquad (18.2)$$

Since each of $r_1, r_2, \ldots, r_{\phi(m)}$ is relatively prime to m, so is their product, and (18.2) may be divided by $r_1 r_2 \ldots r_{\phi(m)}$ which leads to

$$r^{\phi(m)} \equiv 1 \pmod{m}. \qquad (18.3)$$

For example, let $m = 18$; then $\phi(18) = \phi(2) \cdot \phi(9) = 1 \cdot 9(1 - 1/3) = 6$.

Let $r = 5$. Then

$$5^6 \equiv 25^3 \equiv 7^3 \equiv 49 \cdot 7 \equiv 13 \cdot 7 \equiv 91 \equiv 1 \pmod{18}.$$

19. EULER'S PROOF OF THE GENERALIZED THEOREM OF FERMAT

We will base the proof on the so-called Pigeonhole Principle: If n objects are to be placed in fewer than n boxes, then at least one box will have more than one object.

Let now m be a natural number and r an integer relatively prime to m. We consider the numbers

$$r^0 = 1, \ r^1, \ r^2, \ldots, \ r^k \quad \text{where} \quad k = \phi(m).$$

These k + 1 numbers, all of which are relatively prime to m, must be distributed among $k = \phi(m)$ prime residue classes mod m. So at least one prime residue class must contain at least two of the numbers $1, r, \ldots, r^k$; in other words there will exist two integers a and h with $o \leqslant a < a + h \leqslant k = \phi(m)$ such that

$$r^a \equiv r^{a+h} \quad (\text{mod } m) \tag{19.1}$$

Since $(\ r^a, \ m) = 1$, we may divide by r^a; we find

$$1 \equiv r^h \quad (\text{mod } m) \qquad \text{with } h > 0.$$

Consequently there will exist a <u>least positive</u> h, satisfying this congruence, which we designate by e:

$$1 \equiv r^e \quad (\text{mod } m) \quad \text{where } 1 \leqslant e \leqslant k = \phi(m)$$

and also

$$1 \not\equiv r^a \quad (\text{mod } m) \quad \text{if } 1 \leqslant a < e.$$

This e is called the <u>exponent</u> of r mod m, or <u>degree</u> of r mod m. If e > 1, then the numbers $1 = r^0, \ r^1, \ldots, \ r^{e-1}$ are all incongruent mod m. The sequence of powers

$$1, \ r, \ r^2, \ldots, \ r^e, \ r^{e+1}, \ldots \tag{19.2}$$

forms a periodic sequence mod m, that is

$$r^{k_1} \equiv r^{k_2} \ (\text{mod } m) \ \text{if and only if } k_1 \equiv k_2 (\text{mod } e)$$

As we saw, $e \leqslant \phi(m)$. If $\phi(m) > 1$ and $e < \phi(m)$, then there will exist at least one natural number r_1 relatively prime to m which is incongruent to each of the numbers $1, r, r^2, \ldots, r^{e-1}$; hence incongruent to r^n for all $n > 0$. We form the products

$$34 \quad r_1 \cdot 1, \; r_1 \cdot r, \ldots, r_1 \cdot r^{e-1}.$$

(19.3)

Each of these is incongruent mod m with each of the numbers

$1, r, r^2, \ldots, r^{e-1}$, because

$$r_1 \, r^j \equiv r^i \pmod{m} \text{ with } 0 \leqslant j \leqslant e-1 \text{ and } 0 \leqslant i \leqslant e-1$$

implies, since $r^e \equiv 1 \pmod{m}$, that

$$r_1 \, r^j \equiv r^{i+e} \pmod{m}$$

so $\quad r_1 \equiv r^{i+e-j} \pmod{m}$

which contradicts $r_1 \equiv r^n \pmod{m}$ for n a natural number.

In addition, the numbers (19.3) are incongruent among themselves, because, with $0 \leqslant i < j \leqslant e-1$

$$r_1 \, r^i \equiv r_1 \, r^j \pmod{m} \quad \text{implies} \quad r^i \equiv r^j \pmod{m}$$

in contradiction to the above.

Thus there exist 2e numbers

$$1, \, r, \ldots, \, r^{e-1}, \, r_1, \, r_1 \, r, \ldots, \, r_1 \, r^{e-1}$$

which are incongruent mod m and thus $\phi(m) \geqslant 2e$.

If $\phi(m) = 2e$, then $r^{\phi(m)} = r^{2e} = (r^e)^2 \equiv 1^2 = 1 \pmod{m}$.

If $\phi(m) > 2e$, then we may continue the above process, that is, we may find an r_2 relatively prime to m for which the numbers

$$1, r, \ldots, \, r^{e-1}, \, r_1, \, r_1 \, r, \ldots, \, r_1 \, r^{e-1}, \, r_2, \, r_2 \, r, \ldots, r_2 \, r^{e-1}$$

are not only prime to m but also are incongruent to one another mod m. We will then have $\phi(m) \geqslant 3e$;

either $\phi(m) = 3e$ and $r^{\phi(m)} = r^{3e} = (r^e)^3 \equiv 1^3 = 1 \pmod{m}$ or

we will be able to continue the process. But the process must terminate, since $\phi(m)$ is a definite natural number and the relation $ke \leqslant \phi(m)$ implies that $k \leqslant \frac{\phi(m)}{e}$.   Now for the process to terminate, it is necessary for $\phi(m)$ to be some multiple ne of e; but then

$$r^{\phi(m)} = r^{ne} = (r^e)^n \equiv 1^n = 1 \quad (\text{mod } m).$$

<u>COMMENT</u>   The systems of numbers

$$1, r, \ldots, r^{e-1}$$

$$r_1, r_1 r, \ldots, r_1 r^{e-1}$$

$$r_2, r_2 r, \ldots, r_2 r^{e-1}$$

$$\cdot\ \cdot\ \cdot\ \cdot\ \cdot\ \cdot\ \cdot\ \cdot\ \cdot\ \cdot\ \cdot$$

are called cosets of the system of numbers

$$1, r, \ldots, r^{e-1},$$

which we will call the primary coset of r mod m.

From what was proved above we get:

Let $A = \{ a_1, \ldots, a_{\phi(m)} \}$ be a reduced system of residues mod m, r an element of A, e the exponent of r mod m.   Then $e \mid \phi(m)$ and A may partitioned into $\frac{\phi(m)}{e}$ subsets: one in which the elements are congruent to the corresponding elements of the primary coset $\{1, r, \ldots, r^{e-1}\}$ of r and, if $\phi(m) > e$, $\frac{\phi(m)}{e} - 1$ other subsets whose elements are congruent to the corresponding elements of the remaining cosets.

1st <u>EXAMPLE</u>   Let $m = 24$ and $A = \{ 1, 5, 7, 11, 13, 17, 19, 23 \}$.

We calculate the exponent of 5 mod 24

$$5^0 = 1, \ 5^1 = 5, \ 5^2 = 25 \equiv 1 \quad (\text{mod } 24) \text{ so } e = 2$$

and

$$\frac{\phi(24)}{e} = \frac{8}{2} = 4.$$

The set A is partitioned into the subsets (cosets)

$$\{1,5\}; \ \{7, \ 7 \cdot 5 \equiv 11\}, \ \{13, \ 13 \cdot 5 \equiv 17\}, \ \{19, \ 19 \cdot 5 \equiv 23\}.$$

2nd EXAMPLE Let $m = 18$ and $A = \{1, 5, 7, 11, 13, 17\}$.

We calculate the exponent of 5 mod 18:

$$5^0 = 1, \ 5^1 = 5, \ 5^2 = 25 \equiv 7, \ 5^3 \equiv 35 \equiv 17 \equiv -1$$

$$5^4 \equiv -5 \equiv 13, \ 5^5 \equiv -25 \equiv -7 \equiv 11, \ 5^6 \equiv -35 \equiv 1 \quad (\bmod \ 18)$$

so $e = 6$ and $\dfrac{\phi(18)}{6} = \dfrac{6}{6} = 1.$

Thus the numbers of the set A are congruent mod 18 to the corresponding elements of the primary coset

$$\{5^0 \equiv 1, \ 5^1 \equiv 5, \ 5^2 \equiv 7, \ 5^3 \equiv 17, \ 5^4 \equiv 13, \ 5^5 \equiv 11\}.$$

PROBLEMS FOR CHAPTER 2

1. Show by using congruence mod 4 that $x^2+y^2 = m$ is insolvable for $m \equiv 3 \pmod 4$. Hint: Determine the possible values of x^2+y^2 mod 4.

2. Show that $x^2+2y^2 = m$ is not solvable for m if $7|m$ but $49 \nmid m$. Hint: Consider possible values mod 7.

3. a) Show that a positive integer of the form $8b+7$ cannot be decomposed into a sum of three squares of integers;
$$8b+7 \neq x^2+y^2+z^2.$$

   b) Show that a number of the form $4^a(8b+7)$ cannot be decomposed into a sum of three squares. Hint: Show that if $a > 1$ then x, y and z must be even. Then use the method of descent.

4. Note $40 \equiv 1 \pmod{13}$. Use this to construct a divisibility test for 13, analogous to those in §14.

5. Show that the difference of two consecutive cubes is never divisible by 3 or 5.

6. What days of the week will Jan. 1, 1999 and Feb. 28, 2001 fall on?

7. Show that $2^{37}-1$ is divisible by 223.

8. Use the theorem of Fermat to solve
$$7x \equiv 5 \pmod{13}.$$

9. Use section §15 and the theorem of Fermat to solve
$$21x \equiv 24 \pmod{33}.$$

2.P.2

10. Suppose that $(a,m) = 1$. Refer to problem 14 of §1 and find a formula for
the solution x of

$$ax \equiv b \pmod{m}$$

in terms of the bracket symbol there used and the quotients q_i in the
Euclidean algorithm applied to a and m, with $a_1 = m$, $a_2 = a$ and $a_1 = q_1 a_2 + a_3$. (This is virtually identical to the method of continued
fractions in Chapter 3.)

Chapter 3
Linear Congruences

20. THE LINEAR CONGRUENCE AND ITS SOLUTION

We call a <u>linear</u> <u>congruence</u> a congruence of the form

$$ax \equiv b \qquad (\bmod\ m) \tag{20.1}$$

where m is a natural number, a and b given integers, and x an
unknown integer.

A solution of the congruence is an integer x which makes it true.
If an integer x_1 is a solution, then all integers of the form $x_1 + km$,
where $k \in \mathbb{Z}$, are also solutions. These solutions we will not regard as
essentially different; we will count them as just one solution. We will
consider two solutions as different if they are incongruent modulo m.
If $\{r_1, r_2, \ldots, r_m\}$ is a complete system of residues mod m, then the
number of solutions of (20.1) is equal to the number of r_j
$(j = 1, \ldots, m)$ which satisfy (20.1).

For (20.1) we have the following:

If $a \equiv 0 \pmod{m}$, then $ax \equiv b \pmod{m}$ does not have a solution if
$b \not\equiv \pmod{m}$ because $0 \cdot x \equiv 0 \pmod{m}$ for any integer x, while if
$b \equiv 0 \pmod{m}$ then $ax \equiv b \pmod{m}$ has any integer as solution, and so
has m distinct solutions.

Let now $a \not\equiv 0 \bmod m$. We will distinguish two cases.

1st CASE The coefficient a is relatively prime to m; $(a, m) = 1$.

The set $\{ar_1, ar_2, \ldots, ar_m\}$ is simultaneously with $\{r_1, \ldots, r_m\}$,
a complete system of residues mod m. (For there are m elements in the

set, and $ar_i \equiv ar_j \Rightarrow r_i \equiv r_j \pmod{m}$, since $(a, m) = 1$.

The integer b will be congruent with some definite one among the numbers $ar_1,\ldots, ar_m$. Consequently, (20.1) has in this case just one solution.

This may also be determined in the following way. Let θ be a positive integer such that $a^\theta \equiv 1 \pmod{m}$, for example, $\theta = \phi(m)$ or $\theta = e = $ (exponent of a mod m). The number $a^{\theta-1}$ will be relatively prime to m. Hence (20.1) is equivalent to

$$a^{\theta-1}\cdot ax \equiv a^{\theta-1}b \pmod{m}$$

whence

$$a^\theta x \equiv 1\cdot x \equiv a^{\theta-1}b \pmod{m}.$$

Thus (20.1) has only one solution, the residue class of integers congruent to $a^{\theta-1}b$. *

For example: $4x \equiv 1 \pmod{5}$. We have $4 \equiv -1 \pmod{5}$ and $4^2 \equiv (-1)^2 \equiv 1 \pmod{5}$. So the exponent of 4 mod 5 is 2, and $4x \equiv 1 \pmod{5}$ has the unique solution $x \equiv 4^{2-1}\cdot 1 \equiv 4 \pmod{5}$.

2nd CASE $(a, m) = d > 1$.

A necessary and sufficient condition for (20.1) to have a solution is that $b \equiv 0 \pmod{d}$.

Actually, from the existence of a solution ρ follows the equality

$$b = a\rho + mk \qquad \text{with integer } k.$$

Since d is a divisor of m and a, it is also a divisor of b, so $b \equiv 0 \pmod{d}$. On the other hand, let $b \equiv 0 \pmod{d}$.

As is evident, the congruence $ax \equiv b \pmod{m}$ has the same solutions as

$$\frac{a}{d} x \equiv \frac{b}{d} \pmod{\frac{m}{d}}. \tag{20.2}$$

---

* Another computational method for solving such congruences will be presented in §24.

However $\left(\frac{a}{d}, \frac{m}{d}\right) = 1$. So (20.2) has just one solution x_0 mod $\frac{m}{d}$,

that is, it is satisfied by $x = x_0 + h \cdot \frac{m}{d}$ where $h \in \mathbb{Z}$, and only by

these integers. Hence also (20.1) has for solutions the numbers

$x_0 + h \left(\frac{m}{d}\right)$ with $h \in \mathbb{Z}$. Two such solutions $x_0 + h_1 \frac{m}{d}$ and $x_0 + h_2 \frac{m}{d}$

are congruent mod m if and only if $(h_1 - h_2) \frac{m}{d} \equiv 0 \pmod{m}$, which is

equivalent to $\dfrac{h_1 - h_2}{d} \cdot m = km$ with integer k.

But this is equivalent to

$$\frac{h_1 - h_2}{d} = k$$

and then to $h_1 - h_2 = kd$ and finally $h_1 \equiv h_2 \pmod{d}$.

Hence (20.1) has, mod m, the following distinct solutions

$$x_0, \ x_0 + \frac{m}{d}, \ x_0 + 2 \cdot \frac{m}{d}, \ldots, \ x_0 + (d-1) \cdot \frac{m}{d},$$

and the number of solutions is d.

For example, let $5x \equiv 15 \pmod{10}$. We have $(5,10) = 5$ and

$15 \equiv 0 \pmod 5$, so the congruence is solvable and has 5 distinct

solutions mod 10; which are the following

$$\{3, \ 3 + 2 = 5, \ 3 + 2 \cdot 2 = 7, \ 3 + 3 \cdot 2 = 9, \ 3 + 4 \cdot 2 = 11\}.$$

21. SYSTEMS OF LINEAR CONGRUENCES

Let k linear congruences be given

$$a_1 x \equiv b_1 (\bmod\ n_1), \ldots, a_k x \equiv b_k\ (\bmod\ n_k) \tag{21.1}$$

We seek those integers x which saisfy all these congruences

simultaneously. For such a solution to exist, it is evidently necessary

that each of the k congruences (21.1) be solvable. Let us suppose this

to be true; the x being sought then satisfies the k congruences:

$$x \equiv r_1\ (\bmod\ m_1), \ \ldots, \ x \equiv r_k\ (\bmod\ m_k) \tag{21.2}$$

where m_j (j = 1,..., k) are certain suitable divisors of n_j

and $r_1, \ldots, r_k$ are definite integers. Conversely, each solution

of the system (21.2) is a solution of (21.1).

We now observe that each solution of (21.2) has the form

$x = r_1 + ym_1$, with integer y and such that the following con-

gruences hold

$$y\ m_1 \equiv r_2 - r_1\ (\bmod\ m_2), \ldots, y\ m_1 \equiv r_k - r_1\ (\bmod\ m_k). \tag{21.3}$$

Conversely, if an integer y satisfies (21.3), then $x = r_1 + y\ m_1$

will satisfy the system (21.2), so also (21.1). In this way we reduce

(21.1) to the solution of a system of k-1 congruences. Continuing the

process, we will arrive, provided no unsolvable congruence appears,

at the solution of only a single linear congruence.

FIRST EXAMPLE: $2x \equiv 3 \ (\bmod\ 5)$, $3x \equiv 1 \ (\bmod\ 4)$, $5x \equiv 2 \ (\bmod\ 3)$.

Since (2, 5) = (3, 4) = (5, 3) = 1, each congruence has a unique

solution and indeed they are:

$$x \equiv 4 \ (\bmod\ 5), \ \ x \equiv 3 \ (\bmod\ 4), \ \ x \equiv 1 \ (\bmod\ 3).$$

We set, with integer k_1,

$$x = 4 + 5 k_1.$$

We introduce this expression for x into the last two congruences and get, for k_1, the following:

$$4 + 5 k_1 \equiv 3 \pmod 4, \qquad 4 + 5 k_1 \equiv 1 \pmod 3$$

whence

$$k_1 \equiv -1 \pmod 4, \qquad k_1 \equiv 0 \pmod 3. \qquad (21.4)$$

From $k_1 \equiv -1 \pmod 4$ it follows that $k_1 = -1 + 4 k_2$ with integer k_2. We introduce this value into the last congruence of (21.4) and obtain

$$-1 + 4 k_2 \equiv 0 \pmod 3$$

which gives

$$k_2 \equiv 1 \pmod 3.$$

We have found

$$k_2 = 1 + 3k \quad \text{with integer } k$$
$$k_1 = -1 + 4 k_2 = -1 + 4 (1 + 3 k) = 3 + 12 k$$
$$x = 4 + 5 k_1 = 4 + 5 (3 + 12 k) = 19 + 60 k.$$

Hence the solutions of the given system are the integers

$$x = 19 + 60 k \quad \text{with} \quad k \in \mathbf{Z}.$$

Naturally the same solution would have been found had we, for example, started out from the second equation of (21.4), setting $k_1 = 3 k_2$ with integer k_2 and introducing this expression into the first congruence of (21.4). We find

$$3 \, k_2 \equiv -1 \quad (\text{mod } 4)$$

$$k_2 \equiv 1 \quad (\text{mod } 4)$$

$$k_2 = 1 + 4 \, k$$

$$k_1 = 3 \, (1 + 4 \, k) = 3 + 12 \, k$$

and

$$x = 4 + 5 \, k_1 = 4 + 5 \, (3 + 12 \, k) = 19 + 60 \, k.$$

The solutions may be represented as a single solution

$$x \equiv 19 \quad (\text{mod } 60)$$

where 60 is the least common multiple of 5, 4, 3.

22.   THE CASE WHEN THE MODULI $m_1, \ldots, m_k$ OF THE SYSTEM OF CONGRUENCES ARE PAIRWISE RELATIVELY PRIME

We set $M = m_1 m_2 \ldots m_k$ and note that M is also the least common multiple of $m_1, \ldots, m_k$ since they are pairwise relatively prime. Next we set

$$M = m_1 M_1 = m_2 M_2 = \ldots = m_k M_k.$$

In other words

$$M_1 = m_2 \ldots m_k, \; M_2 = m_1 m_3 \ldots m_k, \ldots, M_k = m_1 m_2 \ldots m_{k-1}.$$

We have $(m_j, M_j) = 1$ for $j = 1, \ldots, k$ and, in addition, $m_j \mid M_i$ for $i \neq j$. So there exist integers t_j $(j = 1, \ldots, k)$ such that

$$M_j t_j \equiv 1 \pmod{m_j}. \tag{22.1}$$

These integers t_j are independent of the right sides r_j $(j = 1, \ldots, k)$ of the system (21.2).

We now note that if x is an integer that satisfies (21.2) then

$$M_j t_j x \equiv M_j t_j r_j \pmod{m_j} \qquad (j = 1, \ldots, k)$$

so

$$x \equiv M_j t_j r_j \pmod{m_j} \qquad (j = 1, \ldots, k).$$

Moreover, we know

$$r_i M_i t_i \equiv 0 \pmod{m_j} \qquad \text{for } i \neq j,$$

since $m_j \mid M_i$ for $i \neq j$. Thus

$$x \equiv \sum_{i=1}^{k} r_i M_i t_i \pmod{m_j}$$

and so

$$m_j \mid x - \sum_{i=1}^{k} r_i M_i t_i.$$

Thus the least common multiple M also divides $x - \sum_{i=1}^{k} r_i M_i t_i$ and we have

$$x \equiv \sum_{i=1}^{k} r_i M_i t_i \quad (\text{mod } M).$$

(22.2)

So if the system (21.2) is solvable, there will exist a single solution modulo M, given by (22.2).

Conversely, consider the integer $\sum_{i=1}^{k} r_i M_i t_i$; this integer is congruent modulo m_i to $r_i M_i t_i$ since $m_i \mid M_j$ for $i \neq j$.

We thus have, for $i = 1,\ldots, k$

$$x = \sum_{j=1}^{k} r_j M_j t_j \equiv r_i M_i t_i \equiv r_i \cdot 1 \equiv r_i \quad (\text{mod } m_i).$$

Hence $x = \sum_{j=1}^{k} r_j M_j t_j$ is a solution of (21.2).

Suming up we have the

THEOREM 1 (CHINESE REMAINDER THEOREM) If the moduli of the congruences

$$x \equiv r_i \quad (\text{mod } m_i) \quad i = 1,\ldots,k$$

are pairwise relatiely prime, then the congruences have a solution and the solutions form a residue class mod $M = m_1\ldots m_k$.

The solutions may be determined from (22.1) and (21.2).

23. DECOMPOSITION OF A FRACTION INTO A SUM OF AN INTEGER AND PARTIAL FRACTIONS

Let A be an integer and $M \geqslant 2$ a natural number equal to the product $m_1 \ldots m_k$ where $(m_i, m_j) = 1$ for $i \neq j$. We seek to decompose the fraction A/M into a sum of fractions with denominators $m_1, m_2, \ldots, m_k$, and an integer.

Let

$$A \equiv r_1 \pmod{m_1}, \ldots, A \equiv r_k \pmod{m_k}.$$

Using the symbolism of the previous section:

$$M_j = \frac{M}{m_j} = m_1 \ldots m_{j-1} \, m_{j+1} \ldots m_k, \qquad M_j \, t_j \equiv 1 \pmod{m_j}$$

for $j = 1, \ldots, k$,

we have $A - r_1 \equiv 0 \pmod{m_1}$, so, since $m_1 | M_2, M_3, \ldots, M_k$,

$$A - r_1 M_1 t_1 \equiv A - \sum_{i=1}^{k} r_i M_i t_i \equiv 0 \pmod{m_1}.$$

In the same way we find

$$A - \sum_{i=1}^{k} r_i M_i t_i \equiv 0 \pmod{m_2}, \ldots, A - \sum_{i=1}^{k} r_i M_i t_i \equiv 0 \pmod{m_k}.$$

Since $(m_i, m_j) = 1$ for $i \neq j$, $A - \sum_{i=1}^{k} r_i M_i t_i$ is divisible by the least common multiple $m_1 m_2 \ldots m_k = M$ of $m_1, \ldots, m_k$.

(This also follows immediately from $A \equiv r_i \pmod{m_i}$ if we use the results of §22.) Consequently

$$A = r_1 m_2 \ldots m_k \, t_1 + r_2 m_1 m_3 \ldots m_k \, t_2 + \cdots + r_k m_1 \ldots m_{k-1} \, t_k + gM,$$

where $g \in \mathbb{Z}$. Thus

$$\frac{A}{M} = g + \frac{r_1 t_1}{m_1} + \frac{r_2 t_2}{m_2} + \cdots + \frac{r_k t_k}{m_k}$$

$$= g + g_1 + \frac{a_1}{m_1} + g_2 + \frac{a_2}{m_2} + \cdots + g_k + \frac{a_k}{m_k}$$

with $0 \le a_j < m_j$ for $j = 1,\ldots,k$, and consequently

$$\frac{A}{M} = G + \frac{a_1}{m_1} + \cdots + \frac{a_k}{m_k} \text{ with integer } G. \qquad (23.1)$$

We have succeeded in decomposing $\frac{A}{M}$ in the desired manner.

I say now that this decomposition is unique. Indeed, let

$$\frac{A}{M} = G' + \frac{a_1'}{m_1} + \cdots + \frac{a_k'}{m_k} \text{ with } 0 \le a_j' < m_j \ (j = 1,\ldots,k)$$

and integer G'. Then

$$G' - G = \frac{a_1 - a_1'}{m_1} + \cdots + \frac{a_k - a_k'}{m_k} ;$$

so, multiplying by the integer $M = m_1 \cdots m_k$,

$$(G' - G) M = \text{integer} = (a_1 - a_1') M_1 + (a_2 - a_2') M_2 + \cdots + (a_k - a_k') M_k,$$

where $|a_j - a_j'| < m_j$ for $j = 1,\ldots,k$. We note that m_1 is a divisor

of $(G - G') M$ as well as of $M_2,\ldots, M_k$. Hence m_1 is a divisor of

$(a_1 - a_1') M_1$, so, since m_1 is relatively prime to M_1, m_1 is a divisor

of $a_1 - a_1'$. However, this implies that $a_1 - a_1' = 0$ since

$|a_1 - a_1'| < m_1$. In the same way we find that $a_j - a_j' = 0$ for

$j = 2,\ldots,k$. Consequently, $G - G' = 0$ and the uniqueness is proved.

More specifically, if in the fraction A/M, M is decomposed into a

product of powers of primes

$$M = p_1^{\alpha_1} \cdots p_k^{\alpha_k} \text{ with } \alpha_j \ge 1 \text{ for } j = 1,\ldots, k,$$

we will have

$$\frac{A}{M} = G + \frac{a_1}{p_1^{\alpha_1}} + \cdots + \frac{a_k}{p_k^{\alpha_k}} \text{ where } 0 \le a_j < p_j^{\alpha_j}.$$

The fraction $a_j/p_j^{\alpha_j}$ is equal to the sum

$$\frac{a_j}{p_j^{\alpha_j}} = \frac{z_0^{(j)}}{p_j^{\alpha_j}} + \frac{z_1^{(j)}}{p_j^{\alpha_j-1}} + \cdots + \frac{z_{\alpha_j-1}^{(j)}}{p_j}$$

where $z_0^{(j)}, z_1^{(j)}, \ldots, z_{\alpha_j-1}^{(j)}$ are the "digits" in the representation of a_j in the system with base p_j:

$$a_j = z_0^{(j)} + z_1^{(j)} p_j + z_2^{(j)} p_j^2 + \cdots + z_{\alpha_j-1}^{(j)} p_j^{\alpha_j-1}$$

where $0 \leqslant z_i^{(j)} < p_j$ $(i=0,\ldots, \alpha_j-1)$, and the expansion ends with the term in $p_j^{\alpha_j-1}$ because $0 \leqslant a_j < p_j^{\alpha_j}$.

<u>COMMENT</u> In the decomposition (23.1) of A/M we will have

$(a_j, m_j) = 1$ for $j = 1,\ldots, k$ if $(A,M) = 1$. Indeed we have

$$\frac{r_j t_j}{m_j} = g_j + \frac{a_j}{m_j} ,$$

so $r_j t_j = m_j g_j + a_j$ and consequently $r_j t_j \equiv a_j \pmod{m_j}$.

However, $(r_j, m_j) = 1$ because $A \equiv r_j \pmod{m_j}$ and $(A, m_j) = 1$.

In addition, $(t_j, m_j) = 1$, because $M_j t_j \equiv 1 \pmod{m_j}$. So

$(r_j t_j, m_j) = 1$. Consequently, since $r_j t_j \equiv a_j \pmod{m_j}$,

we will have $(a_j, m_j) = 1$.

<u>EXAMPLE</u> $\qquad \frac{7}{15} = -1 + \frac{2}{3} + \frac{4}{5}$.

For $15 = 3 \cdot 5$ so $m_1 = 3$, $m_2 = 5$, $M_1 = 5$, $M_2 = 3$;

then $t_1 = 2$, $t_2 = 2$. Also

$$r_1 \equiv 7 \equiv 1 \pmod{m_1}, \quad r_2 \equiv 7 \equiv 2 \pmod{m_2}$$

$$A = 7 \equiv 1 \cdot 5 \cdot 2 + 2 \cdot 3 \cdot 2 \pmod{M = 15}$$

$$A = 7 = -1 \cdot 15 + 1 \cdot 5 \cdot 2 + 2 \cdot 3 \cdot 2$$

$$\frac{A}{M} = \frac{7}{15} = -1 + \frac{2}{3} + \frac{4}{5} .$$

3.24.1

24. SOLUTION OF LINEAR CONGRUENCES WITH THE HELP OF CONTINUED FRACTIONS

Consider the congruence $ax \equiv b \pmod{m}$ with $(a, m) = 1$. As we have seen, this congruence has an infinite number of integer solutions all belonging to one residue class modulo m. If x_0 is one of the solutions, then $ax_0 = b + my_0$ with integer y_0. The pair (x_0, y_0) is then a solution of the Diophantine equation $ax - my = b$ with unknown integers x, y.[*] Conversely, if (x_1, y_1) is a solution of this Diophantine equation, that is $ax_1 - my_1 = b$, then $ax_1 = b + my_1$ so $ax_1 \equiv b \pmod{m}$. Thus the solution of the congruence $ax \equiv b \pmod{m}$ may be reduced to the solution of the Diophantine equation $ax - my = b$. For the solution of the Diophantine equation, we expand the corresponding fraction a/m in a continued fraction as follows:

First we apply the Euclidean algorithm to the pair $\{a, m\}$; since $(a, m) = 1$, setting $m = a_1$, there will result, for some suitable n, the following n successive equations of division:

$$a = q_1 a_1 + a_2 \quad \text{with integer } q_1 \text{ and } 0 < a_2 < a_1$$

$$a_1 = q_2 a_2 + a_3 \quad \text{with } q_2 \text{ a natural number and } 0 < a_3 < a_2$$

$$\cdots \cdots \cdots \cdots \cdots \cdots$$

$$a_{n-2} = q_{n-1} a_{n-1} + a_n \quad \text{with } q_{n-1} \text{ a natural number and } 0 < a_n = 1 < a_{n-1}$$

$$a_{n-1} = q_n a_n \quad \text{with } q_n = a_{n-1}$$

since

$$a_n = (a, m) = 1.$$

[*]Called a Diophantine equation after Diophantos (c. 300 AD), whose treatise Arithmetika is partially devoted to the solution of equations in positive rational numbers.

We may rewrite these equations as follows

$$\frac{a}{a_1} = q_1 + \frac{a_2}{a_1} = q_1 + \frac{1}{a_1/a_2}$$

$$\frac{a_1}{a_2} = q_2 + \frac{a_3}{a_2} = q_2 + \frac{1}{a_2/a_3}$$
$$\cdots \cdots \cdots \cdots \cdots \cdots$$
$$\frac{a_{n-2}}{a_{n-1}} = q_{n-1} + \frac{a_n}{a_{n-1}} = q_{n-1} + \frac{1}{a_{n-1}/a_n}$$

$$\frac{a_{n-1}}{a_n} = q_n.$$

Consequently, with elimination of a_1, a_2,..., a_n we find the

following expansion of a/m as a continued fraction

$$\frac{a}{m} = q_1 + \cfrac{1}{q_2 + \cfrac{1}{q_3 + \cfrac{}{\ddots + \cfrac{1}{q_{n-1} + \cfrac{1}{q_n}}}}} \tag{24.1}$$

We now set

$$Z_1 = q_1, \quad N_1 = 1, \quad Z_2 = q_1 q_2 + 1, \quad N_2 = q_2 \quad \text{and}$$

$$\begin{cases} Z_k = q_k Z_{k-1} + Z_{k-2} \\ N_k = q_k N_{k-1} + N_{k-2} \end{cases} \quad \text{for } 3 \leqslant k \leqslant n. \tag{24.2}$$

We have

$$q_1 = \frac{Z_1}{N_1}, \quad q_1 + \frac{1}{q_2} = \frac{q_1 q_2 + 1}{q_2} = \frac{Z_2}{N_2},$$

$$q_1 + \cfrac{1}{q_2 + \cfrac{1}{q_3}} = q_1 + \frac{1}{(q_2 q_3 + 1)/q_3} = q_1 + \frac{q_3}{q_2 q_3 + 1}$$

$$= \frac{q_3(q_1 q_2 + 1) + q_1}{q_2 q_3 + 1} = \frac{q_3 Z_2 + Z_1}{q_3 N_2 + N_1} = \frac{Z_3}{N_3}.$$

We now assume that for some k, $3 \leqslant k \leqslant n-1$, we have

$$q_1 + \cfrac{1}{q_2 + \cfrac{1}{\ddots + \cfrac{1}{q_k}}} = \frac{q_k Z_{k-1} + Z_{k-2}}{q_k N_{k-1} + N_{k-2}} = \frac{Z_k}{N_k} . \qquad (24.3)$$

Then we will have

$$q_1 + \cfrac{1}{q_2 + \cfrac{1}{\ddots + \cfrac{1}{q_k + \cfrac{1}{q_{k+1}}}}} = \frac{(q_k + \frac{1}{q_{k+1}}) Z_{k-1} + Z_{k-2}}{(q_k + \frac{1}{q_{k+1}}) N_{k-1} + N_{k-2}}$$

$$= \frac{(q_k q_{k+1} + 1) Z_{k-1} + q_{k+1} Z_{k-2}}{(q_k q_{k+1} + 1) N_{k-1} + q_{k+1} N_{k-2}}$$

$$= \frac{q_{k+1}(q_k Z_{k-1} + Z_{k-2}) + Z_{k-1}}{q_{k+1}(q_k N_{k-1} + N_{k-2}) + N_{k-1}}$$

$$= \frac{q_{k+1} Z_k + Z_{k-1}}{q_{k+1} N_k + N_{k-1}} = \frac{Z_{k+1}}{N_{k+1}} .$$

Hence, using mathematical induction, we see that (24.3) is true for every k with $3 \leqslant k \leqslant n$.

We now form the expression $Z_k N_{k-1} - Z_{k-1} N_k$ for $3 \leqslant k \leqslant n$. We have

$$\begin{aligned}
Z_k N_{k-1} - Z_{k-1} N_k &= (q_k Z_{k-1} + Z_{k-2}) N_{k-1} - Z_{k-1}(q_k N_{k-1} + N_{k-2}) \\
&= -(Z_{k-1} N_{k-2} - Z_{k-2} N_{k-1}) \\
&= +(Z_{k-2} N_{k-3} - Z_{k-3} N_{k-2}) \\
&\quad \cdot \cdot \cdot \cdot \cdot \cdot \cdot \cdot \cdot \cdot \cdot \cdot \\
&= (-1)^{k-2}(Z_2 N_1 - Z_1 N_2) \\
&= (-1)^{k-2}(q_1 q_2 + 1 - q_1 q_2) = (-1)^{k-2} \cdot 1 = (-1)^{k-2} .
\end{aligned}$$

So we have, for $2 \leqslant k \leqslant n$

$$Z_k N_{k-1} - Z_{k-1} N_k = (-1)^k . \qquad (24.4)$$

But from this we immediately conclude that Z_k and N_k (for $k = 2,\ldots,n$) are relatively prime, and this is also true for $k = 1$: $Z_1 = q_1$, $N_1 = 1$. Additionally, we note that $N_1,\ldots, N_k$ are positive.

From (24.1) and (24.3) it follows that

$a/m = a/a_1 = Z_n/N_n$, and, since the fractions a/m and Z_n/N_n are reduced and $m > 0$, $N_n > 0$, , we have $a = Z_n$, $m = N_n$.

Applying now (24.4) for $k = n$ we see

$$aN_{n-1} - mZ_{n-1} = (-1)^n$$

and thus, multiplying with $(-1)^n$ b,

$$a\cdot(-1)^n b\, N_{n-1} - m\cdot(-1)^n b\, Z_{n-1} = b.$$

Thus we have arrived at a solution to the Diophantine equation

$ax - my = b$, namely

$$x_o = (-1)^n b\, N_{n-1}, \qquad y_o = (-1)^n b\, Z_{n-1}.$$

For any other solution (x, y) of the same Diophantine equation we will have

$$a\,(x-x_o) - m(y-y_o) = 0$$

so $\qquad \dfrac{x-x_o}{m} = \dfrac{y-y_o}{a} = u = \text{integer}^*.$

Consequently

$$x = x_o + mu, \quad y = y_o + au \quad \text{with} \quad u \in \mathbb{Z}.$$

From the above we finally conclude that the congruence

$ax \equiv b \pmod{m}$ has for solutions the integers

$$x = (-1)^n b\, N_{n-1} + mu \text{ with } u \in \mathbb{Z}.$$

---

*since $m \mid a(x-x_o)$ and $(a,m) = 1$, we have $m \mid x-x_o$.

<u>EXAMPLE</u> Let the congruence be $30x \equiv 7 \pmod{37}$

We have

$$\frac{30}{37} = 0 + \cfrac{1}{1 + \cfrac{1}{4 + \cfrac{1}{3 + \cfrac{1}{2}}}}$$

so $q_1 = 0$, $q_2 = 1$, $q_3 = 4$, $q_4 = 3$, $q_5 = 2$

Consequently

$$N_1 = 1 \quad N_2 = q_2 = 1 \quad N_3 = 4 \cdot 1 + 1 = 5 \quad N_4 = 3 \cdot 5 + 1 = 16$$

So a solution of the congruence is

$$x_o = (-1)^5 \cdot 7 \cdot 16 = -112$$

For the Diophantine equation $30x - 37y = 7$, a solution is the pair
$x_o = -112$, $y_o = (-1)^5 \cdot 7 \cdot 13 = -91$ since
$$Z_1 = q_1 = 0 \quad Z_2 = q_1 q_2 + 1 = 1 \quad Z_3 = 4 \cdot 1 + 0 = 4 \quad Z_4 = 3 \cdot 4 + 1 = 13.$$

From this pair we get, with $u = 3$

$$x = x_o + mu = -112 + 37 \cdot 3 = -1$$
$$y = y_o + mu = -91 + 30 \cdot 3 = -1$$

3.P.1

PROBLEMS FOR CHAPTER 3

1. Use the method of §21 to find the solutions (if they exist) of

   a) $x \equiv 11$ (mod 12), $x \equiv 5$ (mod 18), $x \equiv 20$ (mod 21).

   b) $x \equiv 3$ (mod 12), $x \equiv 9$ (mod 18), $x \equiv 1$ (mod 21).

2. Using the notation of §22, set $e_j = t_j M_j$ $j = 1,\ldots,k$.
   Show $e_i e_j \equiv \begin{cases} e_i & \text{for } i = j \\ 0 & \text{for } i \neq j \end{cases}$ (mod M). These e_i are referred to as
   orthogonal idempotents in the theory of rings and algebras.

3. Compute e_1, e_2, e_3 for the case $m_1 = 5$, $m_2 = 8$, $m_3 = 13$.

   a) Using these e_i solve the system

   $$x \equiv 3 \ (\text{mod } 5) \qquad x \equiv 1 \ (\text{mod } 8) \qquad x \equiv 6 \ (\text{mod } 13).$$

   b) Consider the subset of residues $\{0, e_1, 2e_1, 3e_1, 4e_1\}$ (mod M =
       520). Form an addition and multiplication table for this subset.
       What system is this?

   c) Define a correspondence between the residue class a (mod m) and the
       vector (a_1, a_2, a_3) defined by $a \equiv a_i$ (mod m_i), where we take a_i to
       be the least positive residue mod m_i. It is clear that

   $$a \equiv \sum_{i=1}^{3} a_i e_i \ (\text{mod } M = 520).$$

       Let a = 157, b = 139. Find the vectors
       (a_1,a_2,a_3) and (b_1,b_2,b_3) for a and b. Show that a+b and ab
       correspond to the sum and product (termwise) of the vectors, where
       the ith position in the vector is computed mod m_i. Find the inverse
       of $a \equiv 157$ (mod 520) by finding the inverses mod m_i of the vector
       entries.

4. We will now consider problem 3 in full generality. Let $m_1, m_2, \ldots, m_k$ be
   pairwise relatively prime, and $M = m_1 \ldots m_k$. Define the correspondence
   $\alpha \sim (a_1, \ldots, a_k)$ by $a \equiv a_j \pmod{m_j}$, $j = 1, \ldots, k$. Clearly
   $a \equiv \sum_{j=1}^{k} a_j e_j \pmod{M}$. Show that if $a \equiv \sum_{j=1}^{k} x_j e_j$ then $a_j \equiv x_j \pmod{m_j}$,
   $j = 1, \ldots, k$. Use this and problem 2 to show that if $b \sim (b_1, \ldots, b_k)$ then
$$a+b \sim (a_1+b_1, \ldots, a_k+b_k),$$
$$ab \sim (a_1 b_1, \ldots, a_k b_k).$$
   (Readers familiar with modern algebra will recognize this as a
   decomposition of the ring of integers mod M into a direct sum of the
   rings of integers mod m_i, $i = 1, \ldots, k$.)

5. With the notation of problem 4, show that $(a,M) = 1$ if and only if
   $(a_i, m_i) = 1$ for $i = 1, \ldots, k$. Show also that if a_i' is the inverse of a_i
   $\pmod{m_i}$ and a' the inverse of $a \pmod{M}$, assuming these all exist, then
$$a' \sim (a_1', \ldots, a_k').$$

6. Prove the formula for $\phi(m)$ (§11) by using problem 19 of Chapter 1 and
   problem 5 above.

7. Write $\frac{157}{520}$ as a sum of three fractions with denominators 5, 8 and 13.

8. Show that if $m_1, \ldots, m_k$ are pairwise relatively prime and
   $0 < A < M = m_1 \ldots m_k$, then in the unique decomposition
$$\frac{A}{M} = \frac{a_1}{m_1} + \ldots + \frac{a_k}{m_k} - g$$
   with $0 < a_i < m_i$, $i = 1, \ldots, k$ we have $0 \leqslant g < k$.

9. How many numbers in the sequence $1n, 2n, 3n, \ldots, mn$ are multiples of m?

10. Use the method of §24, to find all solutions of

    a) $37x - 29y = 5$,

    b) $517x - 112y = 19$.

11. We will now introduce a slightly more efficient procedure for solving a linear Diophantine equation in two variables; $ax-my = b$ with $(a,m) = 1$. Let $q_1,q_2,\ldots,q_n$ be the quotients in the continued fraction expansion of $\frac{a}{m}$. First note that $Z_j = [q_1,q_2,\ldots,q_j]$ and $N_j = [q_2,q_3,\ldots,q_j]$. Then show that, using Problems 11-14 of Chapter 1
$$x_0 = (-1)^n b[q_{n-1},\ldots,q_2] \text{ and } y_0 = (-1)^n b[q_{n-1},\ldots,q_1].$$
(This algorithm is more efficient because the two brackets may be computed simultaneously.)

12. Use the method of problem 11 to find all solutions of

    a) $53x-89y = 7$,

    b) $414x-293y = 11$.

13. Suppose now that $(a,m) = d$. Show the linear diophantine equation $ax-my = b$ is solvable if and only if $d|b$. Give formulas for the solution as in problem 11.

14. Use the formulas of problem 13 to solve
$$493x-578y = 51.$$

15. Use the fact that $GCD(a,b,c) = ((a,b),c)$ to find a solution to
$$25x+35y+28z = 1.$$

16. Assume that $GCD(a,b,c) = 1$ and $(a,b) = d$. It is clear from problem 15 that a solution to
$$ax+by+cz = 1$$
exists. Find a formula giving an infinite number of solutions in terms of a solution x_0,y_0 of $ax+by = d$, and a solution w_0,z_0 of $dw_0+cz_0 = 1$. The formula need not give all solutions.

17. Suppose $(m_1,m_2) = d$. Find a necessary and sufficient condition for
$$x \equiv a_1 \pmod{m_1}, \quad x \equiv a_2 \pmod{m_2}$$
to be solvable and show the solution is unique mod $LCM(m_1,m_2)$.

3.P.4

The following problems develop some interesting properties of finite symmetric continued fractions.

18. Let $\{q_1, q_2, \ldots, q_n\}$ be the terms in a continued fraction, which then equals

    $Z_n/N_n = [q_1, \ldots, q_n]/[q_2, \ldots, q_n]$. Assume that the q_i have a symmetric pattern:
    $$q_1 = q_n, \quad q_2 = q_{n-1}, \ldots, \quad q_n = q_1.$$

    a) Show $Z_n/N_n = Z_n/Z_{n-1}$ so $Z_{n-1} = N_n$.

    b) Use problem 13 of Chapter 1 to show
    $$Z_n \mid N_n^2 + (-1)^n$$

       so $\quad N_n^2 \equiv (-1)^{n-1} \pmod{Z_n}$.

19. With notations of problem 18, suppose that
    $$Z \mid N^2 + (-1)^a, \quad Z > N > 0, \quad (Z,N) = 1.$$

    Develop Z/N in a continued fraction $\{q_1, q_2, \ldots, q_n\}$ where $n \equiv a \pmod 2$. [A finite continued fraction always has two forms: $\{q_1, \ldots, q_n\}$ with $q_n \geqslant 2$ and $\{q_1, \ldots, q_n-1, 1\}$.] Note the conditions now imply $Z = Z_n$, $N = N_n$, $kZ_n = N_n^2 + (-1)^n$, where k is some integer.

    a) Use problem 13 of Chapter 1 and the last relation above to derive
    $$Z_n \mid N_n - Z_{n-1}.$$

    b) Show $N_n = Z_{n-1}$ by estimating the size of $N_n - Z_{n-1}$ in relation to Z_n, which divides it.

    c) Show, referring to problem 18, that $\{q_1, \ldots, q_n\}$ have a symmetric pattern $q_1 = q_n$, $q_2 = q_{n-1}, \ldots$ Hint: Use uniqueness of continued fraction expansions.

3.P.5

20. With notations of problems 18 and 19, let $Z | N^2 + 1$, $(Z,N) = 1$, $0 < N < Z$, so that Z/N has a symmetric continued fraction with an even number of terms
$$\{q_1, q_2, \ldots, q_k, q_k, \ldots, q_1\}.$$
Show $Z = Z_{k-1}^2 + Z_k^2$ (so Z is a sum of two squares). (Hint: Use problem 15 of Chapter 1.)

Chapter 4
Congruences of Higher Degree

25. GENERALITIES FOR CONGRUENCES OF DEGREE k > 1 AND STUDY
     OF THE CASE OF A PRIME MODULUS

Let

$$f(x) = a_0 x^k + a_1 x^{k-1} + \ldots + a_{k-1} x + a_k \tag{25.1}$$

be a polynomial with integer coefficients a_0, a_1,..., a_k and m a natural
number. We say that the polynomial $f(x)$ is identically congruent to 0 and write

$$f(x) \equiv 0 \quad (\bmod\ m) \text{ identically}$$

if $a_j \equiv 0$ (mod m) for $j = 0,\ldots,k$.

We say that two polynomials $f(x) = \sum_{j=1}^{k} a_j x^{k-j}$

and $g(x) = \sum_{j=0}^{k} b_j x^{k-j}$ are congruent identically modulo m, and write

$$f(x) \equiv g(x) \quad (\bmod\ m) \text{ identically,}$$

if

$$f(x) - g(x) \equiv 0 \quad (\bmod\ m) \text{ identically.}$$

For this, it is necessary and sufficient that the coefficients of equal

powers of x in the two polynomials be congruent: $a_j \equiv b_j$ (mod m) for

$j=0,\ldots,k$. The two congruences $f(x) \equiv 0$ (mod m) and $g(x) \equiv 0$ (mod m)

will then have precisely the same solutions. However, the converse of this

fact is not true: $2x + 3 \equiv 0$ (mod 5) and $4x + 1 \equiv 0$ (mod 5) have the

same solutions 1 + 5u, u $\in$ **Z**, without it being true that

$$2x + 3 \equiv 4x + 1 \quad (\bmod\ 5) \text{ identically.}$$

If in the congruence

$$f(x) = a_0 x^k + \ldots + a_k \equiv 0 \quad (\bmod\ m) \text{ we have } a_0 \not\equiv 0 \ (\bmod\ m), \text{ then the}$$

congruence is said to be <u>of degree k.</u>

4.25.2

We will first occupy ourselves with congruences for which $m = p$, a prime number. We take the polynomial (25.1) and form the difference $f(x) - f(x_1)$ for two different integers x and x_1. We find

$$
\begin{aligned}
f(x) - f(x_1) &= a_0(x^k - x_1^{\,k}) + a_1(x^{k-1} - x_1^{\,k-1}) + \ldots + a_{k-1}(x - x_1) \\
&= (x - x_1)\,\{a_0(x^{k-1} + x^{k-2}x_1 + \ldots + x_1^{\,k-1}) + \ldots \\
&\qquad\qquad + a_{k-2}(x + x_1) + a_{k-1}\} \\
&= (x - x_1)\,g(x)
\end{aligned}
$$

where $g(x) = b_0 x^{k-1} + \ldots + b_{k-1}$ with $b_0 = a_0 \not\equiv 0 \pmod p$.

I) Now let us assume that x_1 is a solution of the congruence of degree k: $f(x) \equiv 0 \pmod p$. Then $-f(x_1)$ is an integer divisible by p and thus the polynomial $f(x) - f(x_1)$ will be congruent mod p identically to $f(x)$, so

$$f(x) \equiv (x - x_1)\,g(x) \pmod p \quad \text{identically with } g(x) \text{ of degree } k-1.$$

II) Let us assume now that the congruence $f(x) \equiv 0 \pmod p$ of degree $k > 1$ has the incongruent solutions $x_1, \ldots, x_s$ (with $1 < s < k$). Then we have

$$f(x) \equiv (x - x_1)\ldots(x - x_s)\,g_s(x) \pmod p \quad \text{identically with}$$

$g_s(x) = c_0 x^{k-s} + c_1 x^{k-s-1} + \ldots + c_{k-s}$ with $c_0 \not\equiv 0 \pmod p$. We know, this holds for $s = 1$. Using mathematical induction we will prove it for each $s < k$. To this purpose, assume it holds for $s-1$ incongruent solutions of a congruence $g(x) \equiv 0 \pmod p$.

We have, according to the preceding,

$$f(x) \equiv (x - x_1)\,g_1(x) \pmod p \quad \text{identically,}$$

with $g_1(x)$ of degree $k-1$. Substituting $x = x_i$ ($i = 2, \ldots, s$) into this congruence we find

$$f(x_i) \equiv (x_i - x_1)\,g_1(x_i) \pmod p.$$

Since $f(x_i) \equiv 0 \pmod p$, p will be a divisor of $(x_i - x_1)\,g_1(x_i)$; but $x_i - x_1 \not\equiv 0 \pmod p$ so $g_1(x_i) \equiv 0 \pmod p$ for $i = 2, \ldots, s$.

Thus $g_1(x) \equiv 0$ (mod p) has the s-1 solutions $x_2,\ldots,x_s$.

Now according to the induction assumption, we have

$$g_1(x) \equiv (x-x_2)\ldots(x-x_s)\, g_s(x) \pmod p \text{ identically and thus}$$

$$f(x) \equiv (x-x_1\,(x-x_2)\ldots(x-x_s)\, g_s(x) \pmod p \text{ identically}$$

which is what we wished to prove. We note also that $g_s(x)$ has degree

$k-1 - (s-1) = k-s$.

III) Consider now the congruence $f(x) \equiv 0$ (mod p) of degree k. It has at

most k incongruent solutions mod p.

Indeed, according to the preceding we have $f(x) \equiv a_0(x-x_1)\ldots(x-x_k)$ (mod p)

identically, if $x_1,\ldots,x_k$ are incongruent solutions of $f(x) \equiv 0$ (mod p).

Let x_{k+1} be incongruent to any of $x_1,\ldots,x_k$, that is

$x_{k+1} - x_j \not\equiv 0$ (mod p) for $j = 1,\ldots,k$. We will have

$$f(x_{k+1}) \equiv a_0(x_{k+1}- x_1)\ldots(x_{k+1}- x_k) \pmod p.$$

All of the factors on the right side are not congruent to 0 mod p, so

$f(x_{k+1}) \not\equiv 0$ (mod p).

IV) Let $f(x)$ be of degree k and $f(x) \equiv 0$ (mod p) have k solutions and

$f(x) \equiv f_1(x)\, f_2(x)$ (mod p) identically. Then the number of incongruent

solutions of $f_1(x) \equiv 0$ (mod p) is equal to its degree and similarly for $f_2(x)$.

<u>Proof.</u> Let $f_1(x) = b_0 x^\ell +\ldots+ b_\ell$ and $f_2(x) = c_0 x^m +\ldots+c_m$. with

$b_0 \not\equiv 0$ and $c_0 \not\equiv 0$ (mod p). Then we will have

$$f(x) \equiv b_0\, c_0 x^{\ell+m}+\ldots+ b_\ell\, c_m \pmod p \text{ identically. Thus } \ell + m = k.$$

Each solution of $f(x) \equiv 0$ (mod p) will be a solution of at least one of the

congruences

$$f_1(x) \equiv 0 \pmod p \text{ or } f_2(x) \equiv 0 \pmod p.$$

Conversely, each solution of one of these congruences will be a solution of

$f(x) \equiv 0$ (mod p). Now if the number of incongruent solutions of

of $f_1(x) \equiv 0$ (mod p) or $f_2(x) \equiv 0$ (mod p) were less than, respectively,

4.25.4

ℓ or m, then the number of solutions of $f(x) \equiv 0$ (mod p) would be less than $\ell + m = k$, which is contrary to our hypothesis. Thus $f_1(x) \equiv 0$ (mod p) must have ℓ solutions and $f_2(x) \equiv 0$ (mod p) must have m solutions, as we wished to prove.

V) An Application. The congruence $x^{p-1} - 1 \equiv 0$ (mod p) has, according to the theorem of Fermat, $p-1$ solutions. Now let $\delta | p-1$, so that $p-1 = \delta q$. We have

$$x^{p-1} - 1 = (x^{\delta} - 1)(x^{\delta(q-1)} + x^{\delta(q-2)} + \ldots + x^{\delta} + 1)$$

where the congruences $x^{\delta} - 1 \equiv 0$ (mod p) and $x^{\delta(q-1)} + \ldots + x^{\delta} + 1 \equiv 0$ (mod p) have respectively, the degrees δ and $\delta(q-1)$. Consequently

$$x^{p-1} - 1 \equiv (x^{\delta} - 1) g(x) \quad \text{(mod } p) \text{ identically,}$$

and, according to IV), the congruence $x^{\delta} - 1 \equiv 0$ (mod p) has δ incongruent solutions mod p.

For example, with $p = 7$ and $\delta = 3 | 7-1$, the congruence $x^3 - 1 \equiv 0$ (mod 7) has the 3 solutions 1, 2, 4.

26. THEOREM OF WILSON

THEOREM If p is a prime then $(p-1)! \equiv -1 \pmod p$. Conversely, if for any natural number n we have $(n-1)! \equiv -1 \pmod n$ then n is a prime number.

Proof of necessity. The congruence $x^{p-1}-1 \equiv 0 \pmod p$ has the p-1 incongruent solutions 1, 2,..., p-1. So according to II) of the preceding section for p > 2 we have $x^{p-1}-1 \equiv (x-1)(x-2)...(x-(p-1)) \pmod p$ identically. For x = 0 we then find

$$- 1 \equiv (-1)(-2)...(- (p-1)) \pmod p.$$

If p is odd, p-1 is even and consequently the last congruence becomes

$$-1 \equiv (-1)^{p-1} 1 \cdot 2 \cdots (p-1) \equiv 1 \cdot 2 \cdots (p-1) \pmod p.$$

If p is even, that is, p = 2, then $-1 \equiv 1 \equiv 1! \equiv (2-1)! \mod 2$. The result is now proved in all cases.

Proof of sufficiency. We assume for some natural number $n \geqslant 3$ we have $(n-1)! \equiv -1 \pmod n$, and we must show that n is a prime number. From the hypothesis it follows that

$$n \mid (n-1)! + 1.$$

If n were not a prime number then there would exist a prime p dividing n where $2 \leqslant p \leqslant n-1$. Consequently, p would divide $(n-1)!$. From $p \mid n$ and $n \mid (n-1)! + 1$ would follow $p \mid (n-1)! + 1$. But from $p \mid (n-1)!$ and $p \mid (n-1)! + 1$ would follow $p \mid (n-1)! + 1 - (n-1)! = 1$, which is absurd. Hence n must be prime.

27. THE SYSTEM $(r, r^2, \ldots, r^\delta)$ OF INCONGRUENT POWERS MODULO A PRIME p

Let p be a prime and r an integer not divisible by p, so $(r, p) = 1$.
Let δ be the exponent of r modulo p, (see §19). We will say that r $\underline{\text{belongs}}$
to the exponent δ mod p. As we know, $\delta \,|\, \phi(p) = p-1$. The powers

$$r, \ r^2, \ldots, \ r^\delta \equiv 1 \pmod{p} \tag{27.1}$$

will be all incongruent mod p. Moreover, we will have, with integer
α, $1 \leqslant \alpha \leqslant \delta$,

$$(r^\alpha)^\delta = r^{\alpha\delta} = (r^\delta)^\alpha \equiv 1^\alpha \equiv 1 \pmod{p}.$$

Thus the δ integers (27.1) are solutions of $x^\delta - 1 \equiv 0 \pmod{p}$ which, as we
know, has precisely δ incongruent solutions. Consequently, the integers (27.1)
form a complete system of incongruent solutions of $x^\delta - 1 \equiv 0 \pmod{p}$.
From this it follows that each integer s with $(s,p) = 1$ which belongs to the
exponent δ, since it satisfies $s^\delta - 1 \equiv 0 \pmod{p}$, will be congruent mod p
to one of the numbers (27.1). Hence we arrive at the question: How many of the
integers (27.1) belong to the exponent δ ? We will show that there are $\phi(\delta)$
such integers. Actually, r^α $(1 \leqslant \alpha \leqslant \delta)$ belongs to the exponent δ if and only
if $(\alpha, \delta) = 1$. For, if $(\alpha, \delta) = \delta_1 > 1$, then

$$(r^\alpha)^{\delta/\delta_1} = r^{(\alpha/\delta_1)\delta} = (r^\delta)^{\alpha/\delta_1} \equiv 1^{\alpha/\delta_1} \equiv 1 \pmod{p}.$$

Therefore r^α belongs to an exponent $\delta_2 \leqslant \delta/\delta_1 < \delta$.* On the other hand,
if $(\alpha, \delta) = 1$,
then the power $(r^\alpha)^k$ of r^α will be congruent to 1 mod p if and only if $\delta \,|\, \alpha k$,
which, since $(\alpha, \delta) = 1$, only occurs when $\delta \,|\, k$. Consequently, the least
positive k for which $(r^\alpha)^k \equiv 1 \pmod{p}$ is δ, and so r^α with $(\alpha, \delta) = 1$
belongs to the exponent δ mod p.

We have proved the following: if an r with $(r,p) = 1$ belongs to the

---

*In fact, r^α belongs to δ/δ_1, see problems.

exponent δ , a divisor of p-1, then there exist, in the case

$\phi(\delta) > 1$, $\phi(\delta)$ -1 other integers incongruent to r and to one another that

belong to the same exponent δ. Moreover, an integer s not congruent to a

power of r cannot belong to δ, since the δ elements (27.1) form a complete

system of incongruent solutions mod p of $x^{\delta} \equiv 1$ (mod p). Hence no more than

$\phi(\delta)$ incongruent integers can belong to δ mod p.

We will now show that there indeed exist $\phi(\delta)$ incongruent integers mod p

which belong to an exponent δ which divides p-1. Actually, let us designate by

$\psi(\delta)$ the yet to be determined number of incongruent integers r which belong to

an exponent δ. Then $\psi(\delta) = \phi(\delta)$ if there exists an r belonging

to δ or $\psi(\delta) = 0$ if there exists no r belonging to δ.

Thus $\phi(\delta) - \psi(\delta) \geqslant 0$. Now we have, since each integer 1, 2,..., p-1 belongs to

some exponent δ which must divide p-1,

$$\sum_{\delta|p-1} \psi(\delta) = p-1.$$

However, we know (Theorem 3 of §11) that

$$\sum_{\delta|p-1} \phi(\delta) = p-1.$$

So

$$\sum_{\delta|p-1} [\phi(\delta) - \psi(\delta)] = 0$$

and, since $\phi(\delta) - \psi(\delta) \geqslant 0$, we must have $\phi(\delta) - \psi(\delta) = 0$ for all δ dividing

p-1, as was to be proved.

EXAMPLE Let p=7. Then p-1=6 has divisors 1, 2, 3, 6.

$\phi(1) = 1$; among the numbers 1, 2, 3, 4, 5, 6 one belongs to 1: 1.

$\phi(2) = 1$; among the numbers 1, 2, 3, 4, 5, 6 one belongs to 2: 6.

$\phi(3) = 2$; among the numbers 1, 2, 3, 4, 5, 6 two belong to 3: 2, 4.

$\phi(6) = 2$; among the numbers 1, 2, 3, 4, 5, 6 two belong to 6: 3, 5.

4.28.1

28 INDICES

Let $p > 2$ be a prime. There exist $\phi(p-1)$ integers in the set $\{1, 2,..., p-1\}$ which belong to the exponent $p-1$. These integers are called <u>primitive roots</u> of p. For example, 3 and 5 are primitive roots of 7.

Let g be a primitive root of p. The system of numbers

$$\{g, g^2,..., g^{p-1}\} \tag{28.1}$$

is a reduced system of residues modulo p. We have, with integer α, $g^\alpha \equiv 1 \pmod{p}$ if and only if $\alpha \equiv 0 \pmod{p-1}$. From this it follows that $g^\alpha \equiv g^\beta \pmod{p}$ if and only if $\alpha \equiv \beta \pmod{p-1}$. The least positive residues of $g, g^2,..., g^{p-1}$ are the numbers $1, 2,..., p-1$ taken in a suitable order. If $g^\alpha \equiv A \pmod{p}$ with $1 \leqslant A \leqslant p-1$ and $1 \leqslant \alpha \leqslant p-1$ then the exponent α is called the <u>index</u> of each element of the residue class which A represents; that is α is called the index of A and of each $A' \equiv A \pmod{p}$. The integer A is called the anti-index[*] of α with respect to g. We will denote the index of A by Ind A and the anti-index of α by Num α. (Ind = Latin Index, Num = Latin Numerus.)

For example, let p = 13. It is easy to determine that g = 2 is a primitive root of 13. Actually, the powers $g^k = 2^k$ are, modulo 13, respectively congruent to

g^k	2^1	2^2	2^3	2^4	2^5	2^6	2^7	2^8	2^9	2^{10}	2^{11}	2^{12}
$\equiv$ (mod 13)	2	4	8	3	6	12	11	9	5	10	7	1

[*]There is no commonly accepted English term for this concept.

4.28.2

From this table we derive the following tables

Num = anti-index	1	2	3	4	5	6	7	8	9	10	11	12
Ind = index	12	1	4	2	9	5	11	3	8	10	7	6

Ind = index	1	2	3	4	5	6	7	8	9	10	11	12
Num = anti-index	2	4	8	3	6	12	11	9	5	10	7	1

For the indices and anti-indices with respect to a primitive root g of the prime $p > 2$ we have the following propositions.

I)   Ind A = Ind B if and only if $A \equiv B \pmod{p}$.

II)   Ind $AB \equiv$ Ind A + Ind $B \pmod{p-1}$.

Actually, from $A \equiv g^{\alpha}$ and $B \equiv g^{\beta} \pmod{p}$ follows

$$AB \equiv g^{\alpha+\beta} \pmod{p}$$

so   Ind $AB \equiv \alpha + \beta =$ Ind A + Ind B.

II´) More generally
$$\text{Ind} \prod_{j=1}^{k} A_j \equiv \sum_{j=1}^{k} \text{Ind } A_j \pmod{p-1}.$$

II´´)   Ind $A^k \equiv k$ Ind $A \pmod{p-1}$.

III) $\alpha)$ Ind $1 = p-1$,   since $g^{p-1} \equiv 1 \pmod{p}$,

$\beta)$ Ind $g = 1$,   since $g^1 \equiv g \pmod{p}$,

$\gamma)$ Ind $-1 =$ Ind $(p-1) = \dfrac{p-1}{2}$.

Indeed, we have $(g^{p-1} - 1) \equiv 0 \pmod{p}$, so
$$(g^{(p-1)/2} - 1)(g^{(p-1)/2} + 1) \equiv 0 \pmod{p}.$$

However, $g^{(p-1)/2} \not\equiv 1 \pmod{p}$ because g belongs to the exponent $p-1$. So we must have $g^{(p-1)/2} + 1 \equiv 0 \pmod{p}$

and thus

$$g^{(p-1)/2} \equiv -1 \pmod{p},$$

which is what we wished to prove.

<u>APPLICATIONS</u> 1) To find the least positive residue congruent to 7^{10^6} mod 13.

We have
$$2^{\text{Ind } 7^{10^6}} \equiv 7^{10^6} \pmod{13} \quad \text{and} \quad 7^{10^6} \equiv v \pmod{13},$$

so $\quad 2^{\text{Ind } 7^{10^6}} \equiv v \pmod{13}$

and thus
$$\text{Ind } 7^{10^6} \equiv \text{Ind } v \pmod{12}.$$

Calculating, we have
$$\text{Ind } 7^{10^6} \equiv 10^6 \cdot \text{Ind } 7 \equiv (-2)^6 \cdot 11 \pmod{12}$$
$$\equiv 4 \cdot 11 \equiv 8 \pmod{12},$$
and thus
$$\text{Ind } 7^{10^6} \equiv 8 \pmod{12}$$

and, using the anti-index table,
$$7^{10^6} \equiv \text{Num } 8 \equiv 9 \pmod{13}.$$
So the desired residue of 7^{10^6} mod 13 is 9.

2) Consider the congruence $ax \equiv b \pmod{p}$ with $p > 2$ prime and $(a,p) = 1$.

If $b \equiv 0 \pmod{p}$ then the solutions are kp with $k \in \mathbb{Z}$. Let now

$(b, p) = 1$. With respect to some primitive root g of p we will have
$$\text{Ind } ax \equiv \text{Ind } a + \text{Ind } x \equiv \text{Ind } b \pmod{p-1},$$

so $\quad \text{Ind } x \equiv \text{Ind } b - \text{Ind } a \pmod{p-1}$.

From Ind x we find the corresponding x.

For example, consider $3x \equiv 2 \pmod{13}$. We take $g = 2$. Then
$$\text{Ind } x \equiv \text{Ind } 2 - \text{Ind } 3 \equiv 1-4 \equiv -3 \equiv 9 \pmod{12}.$$

Hence $x \equiv \text{Num } 9 \equiv 5 \pmod{13}$, and the solutions are $5 + 13k$

with $k \in \mathbb{Z}$.

4.29.1

29. BINOMIAL CONGRUENCES

We call a congruence of the form $ax^n \equiv b \pmod{m}$ a binomial congruence.

Here we will examine the case

$$ax^n \equiv b \pmod{p} \text{ where } p \text{ is an odd prime and } p \nmid a. \tag{29.1}$$

We determine an a' such that

$$aa' \equiv 1 \pmod{p};$$

naturally $p \nmid a'$. The congruence (29.1) is equivalent to (that is, has the same solutions as) $x^n \equiv ba' \pmod{p}$. Thus it suffices for us to concern ourselves with congruences of the form

$$x^n \equiv \Delta \pmod{p}.$$

If $\Delta \equiv 0 \pmod{p}$ then the solutions are $x = kp$ with $k \in \mathbf{Z}$.

Let now $(\Delta, p) = 1$. For solutions of $x^n \equiv \Delta \pmod{p}$ to exist it is necessary and sufficient that with respect to any primitive root g of p we have

$$n \operatorname{Ind} x \equiv \operatorname{Ind} \Delta \pmod{p-1}. \tag{29.2}$$

We distinguish two cases, setting $(n, p-1) = \delta$.

1st Case: $\operatorname{Ind} \Delta \not\equiv 0 \pmod{\delta}$.

Then (29.2) and therefore also (29.1) have no solutions.

2nd Case: $\operatorname{Ind} \Delta \equiv 0 \pmod{\delta}$.

Then (29.2) has δ solutions which are incongruent mod $p-1$, so (29.1) has δ incongruent solutions mod p.

The condition of solvbility $\operatorname{Ind} \Delta \equiv 0 \pmod{\delta}$ may be restated as follows, using $(n, p-1) = \delta$:

$$\operatorname{Ind} \Delta \equiv 0 \pmod{\delta} \iff \frac{p-1}{\delta} \operatorname{Ind} \Delta \equiv 0 \pmod{\frac{p-1}{\delta}\delta}$$
$$\iff \operatorname{Ind} \Delta^{(p-1)/\delta} \equiv 0 \pmod{p-1}$$
$$\iff \Delta^{(p-1)/\delta} \equiv 1 \pmod{p}.$$

EXAMPLES

1) $3x^6 \equiv 5$ (mod 13). We have $3 \cdot (-4) \equiv 1$ (mod 13),

so $3x^6 \equiv 5$ (mod 13) is equivalent to

$x^6 \equiv -4 \cdot 5 \equiv -20 \equiv 6$ (mod 13).

We have $\Delta = 6$ and $\delta = (n, p-1) = (6,12) = 6$. However, with $g = 2$,

Ind $6 = 5 \not\equiv 0$ (mod 6). Consequently, $3x^6 \equiv 5$ (mod 13) has no solutions.

2) $7x^5 \equiv 2$ (mod 13). This time we have $n = 5$, $p = 13$,

$\delta = (n, p-1) = (5, 12) = 1$ so the congruence will be solvable.

$2 \cdot 7 \equiv 1$ (mod 13), so $7x^5 \equiv 2$ (mod 13) is equivalent to $x^5 \equiv 4$ (mod 13).

Thus $\Delta = 4$ and with $g=2$, Ind $4 = 2 \equiv 0$ (mod $\delta = 1$). We find the solutions

by solving 5 Ind $x \equiv$ Ind $4 \equiv 2$ (mod 12).

Since $5 \cdot 5 \equiv 25 \equiv 1$ (mod 12).

$$\text{Ind } x \equiv 5 \cdot 2 \text{ (mod 12)}$$
$$\equiv 10 \text{ (mod 12)}.$$

So

$$x \equiv 10 \text{ (mod 13)}.$$

Hence $7x^5 \equiv 2$ (mod 13) has the solutions $10 + 13k$ with $k \in \mathbb{Z}$.

3) $7x^4 \equiv 11$ (mod 13)   Since $2 \cdot 7 \equiv 1$ (mod 13), this is equivalent to

$x^4 \equiv 2 \cdot 11 \equiv 22 \equiv 9$ (mod 13). Thus $\Delta = 9$ and $\delta = (n,p-1) = (4,12) = 4$.

Since $4 \mid$ Ind $9 = 8$, there will be four solutions. We find then by solving

$$4 \text{ Ind } x \equiv 8 \text{ (mod 12)},$$
$$\text{Ind } x \equiv 2 \text{ (mod 3)}$$
$$\equiv 2, 5, 8, 11 \text{ (mod 12)},$$

which gives the solutions

$$x \equiv 4, 6, 9, 7 \text{ (mod 13)}.$$

30. RESIDUES OF POWERS MOD p.

An integer Δ is called an nth power residue mod m (n and m natural numbers) when the congruence

$$x^n \equiv \Delta \quad (\text{mod } m)$$

has solutions. Any integer $\Delta' \equiv \Delta$ (mod m) is also an nth power residue along with Δ, so all the elements of the residue class which are represented by Δ are nth power residues.

For example the numbers of the residue class containing 0 are nth power residues for any n and m, because $x^n \equiv 0$ (mod m) is solvable since $0^n \equiv 0$ (mod m). Similarly the integer $\Delta = 1$ is an nth power residue modulo m for any n and m because $x^n \equiv 1$ (mod m) is solvable since $1^n \equiv 1$ (mod m).

Below we will concern ourselves more specifically with nth power residues mod p which are relatively prime to a prime number p > 2. In other words, we will study which of the numbers 1, 2,..., p-1 are nth power residues mod p. Put another way, for which of the numbers 1, 2,..., p-1 is the congruence $x^n \equiv \Delta$ (mod p) solvable, and how many incongruent solutions has it?

For the congruence $x^n \equiv \Delta$ (mod p) with $(\Delta, p) = 1$ to be solvable it is necessary and sufficient, as we saw in the previous section, that

$$\Delta^{(p-1)/\delta} \equiv 1 \quad (\text{mod } p) \text{ where } \delta = (n, p-1).$$

Thus we have

<u>THEOREM</u> 1 The integer Δ with $(\Delta, p) = 1$ is an

nth power residue mod p when $\Delta^{\frac{p-1}{\delta}} \equiv 1$ (mod p),

nth power non-residue mod p when $\Delta^{\frac{p-1}{\delta}} \not\equiv 1$ (mod p).

We pose the question: How many incongruent nth power residues mod p are there? The answer comes immediately from the following Theorem.

<u>THEOREM</u> 2 The number of incongruent n^{th} power residues mod p (which are prime to p) is identical with the number of incongruent solutions of

$$x^{\frac{p-1}{\delta}} \equiv 1 \quad (\text{mod } p), \text{ where } \delta = (n, p-1);$$

so (according to application V) of §25), it is equal to $\frac{p-1}{\delta}$.

The proof is immediate.

<u>THEOREM</u> 3 In the system

$$\{r_1^n, r_2^n, \ldots, r_{p-1}^n\} \tag{30.1}$$

where $r_1, \ldots, r_{p-1}$ is a reduced system of residues mod p, appear all the nth power residues mod p which are prime to p, and each one appears δ times.

<u>Proof</u> Indeed the system (30.1) is formed exclusively of nth power residues mod p which are prime to p. Conversely, each such residue is congruent with at least one of the elements of (30.1). As we saw, the number of such incongruent n^{th} power residues is $(p-1)/\delta$. Moreover, each of them appears

$$\frac{p-1}{\frac{p-1}{\delta}} = \delta$$

times, since, as we saw in §29, the congruence $x^n \equiv \Delta \pmod{p}$, when it is solvable, has δ incongruent solutions.

<u>EXAMPLE</u> We consider the congruence $x^n \equiv \Delta \pmod{p}$ with $n = 3$ and $p = 7$; consequently $\delta = (n, p-1) = (3,6) = 3$ and $\Delta = 1, 2, 3, 4, 5, 6$. We must calculate $\Delta^{(p-1)/\delta} = \Delta^{(7-1)/3} = \Delta^2$ for $\Delta = 1, 2, \ldots, 6$. We find

Δ	1	2	3	4	5	6
$\Delta^{(p-1)/\delta} = \Delta^2$	1	4	9	16	25	36
$\Delta^2 \equiv \pmod{7}$	1	4	2	2	4	1 .

Consequently, there are two 3rd power residues mod 7, the integers

1 and 6. The congruence $x^3 \equiv 1$ (mod 7) has the three solutions

1, 2, 4; and the congruence $x^3 \equiv 6$ (mod 7) has the solutions

x = 3, 5, 6. Finally we form the system (30.1)

$$\{r_1, \ldots, r_{p-1}\} = \{1, 2, 3, 4, 5, 6\};$$

We have $\{1^3 \equiv 1,\ 2^3 \equiv 1,\ 3^3 \equiv 6,\ 4^3 \equiv 1,\ 5^3 \equiv 6,\ 6^3 \equiv 6\}$ (mod 7)

from which we see that the two 3rd power residues each occur

$\delta = 3$ times.

31. PERIODIC DECADIC EXPANSIONS

Let a real number A have the periodic decimal expansion

$$A = 0.\, z_1 z_2 \ldots z_k z_1 z_2 \ldots z_k z_1 z_2 \ldots z_k \ldots = 0.\overline{z_1 z_2 \ldots z_k}$$

with period of length k, where the z_i (i=1,..., k) are digits;

$z_i \epsilon \{0, 1, \ldots, 9\}$. Then we have

$$A = \frac{z_1 \ldots z_k}{10^k} + \frac{z_1 \ldots z_k}{10^{2k}} + \frac{z_1 \ldots z_k}{10^{3k}} + \ldots$$

which we recognize as a convergent geometric series, since the multiplier

is $10^{-k} < 1$. Next we have

$$\frac{A}{10^k} = \frac{z_1 \ldots z_k}{10^{2k}} + \frac{z_1 \ldots z_k}{10^{3k}} + \ldots$$

and consequently

$$A \left(1 - \frac{1}{10^k}\right) = \frac{z_1 \ldots z_k}{10^k}.$$

Hence A is equal to a rational number which we will represent by the reduced

fraction z/n, so that (z, n) = 1.

From the equality

$$A = \frac{z_1 \ldots z_k}{10^k - 1} = \frac{z}{n}$$

and the fact that $(z,n) = 1$ we conclude that 10^k-1 is a multiple of n[*],
from which we get $10^k \equiv 1$ (mod n). Thus $(10, n) = 1$. Now let 10 belong to
the exponent e mod n. Then we also have $k \equiv 0$ (mod e). Conversely,
let $z/n < 1$ be a positive rational number written in reduced form;
$(z,n) = 1$, $(n,10) = 1$. We assume 10 belongs to the exponent
k (mod n); that is $10^k \equiv 1$ (mod n) and $10^\ell \not\equiv 1$ (mod n) for $0 < \ell < k$.
Then $(10^k-1)/n$ will be a natural number and the same will be true of the
product $z \cdot [(10^k-1)/n]$. Moreover, since $z/n < 1$, the integer
$z \cdot [(10^k-1)/n] < 10^k-1$, so it may be written in the decimal system
with k digits:

$$\frac{z(10^k-1)}{n} = z_1 z_2 \ldots z_k .$$

Consequently we will have

$$\frac{z}{n} = \frac{z_1 \ldots z_k}{10^k-1} = \frac{z_1 \ldots z_k}{10^k} \cdot \frac{1}{1 - \dfrac{1}{10^k}}$$

$$= \frac{z_1 \ldots z_k}{10^k} \left(1 + \frac{1}{10^k} + \frac{1}{10^{2k}} + \cdots \right)$$

$$= \frac{z_1 \ldots z_k}{10^k} + \frac{z_1 \ldots z_k}{10^{2k}} + \frac{z_1 \ldots z_k}{10^{3k}} + \cdots$$

$$= 0. \overline{z_1 \ldots z_k} .$$

Thus the rational number z/n has a periodic decimal expansion of length k.
I say that this period of length k is primitive, that is, z/n cannot be
written with a decimal expansion of shorter period. Indeed, if there
existed a shorter period $k' < k$, then according to what we saw before,
we would have $10^{k'} \equiv 1$ (mod n) contradicting our hypothesis that
10 belongs to the exponent k mod n.

[*]In fact, $n(z_1 \ldots z_k) = z(10^k-1)$. Since $n | z(10^k-1)$ and $(n, z) = 1$,
we have $n | 10^k-1$.

4.31.2

<u>REMARK</u> If in the decimal expansion $A = 0.\overline{z_1 \ldots z_k} = z/n$ with $(z, n) = 1$ the period $z_1 \ldots z_k$ is primitive, then we have not only $e \mid k$ but $e = k$, where e is the exponent to which 10 belongs mod n. This follows easily from the observations immediately preceding.

<u>EXAMPLES</u> Consider 8/13. Here 10 belongs to the exponent 6 mod 13; so the primitive period of the decimal expansion of 8/13 has length six. Actually,

$$8/13 = 0.\overline{615384}.$$

Now consider $7/33 = .\overline{21}$. Here 10 belongs to the exponent 2 mod 33, since $10^2 \equiv 100 \equiv 100 - 99 \equiv 1 \pmod{33}$.

4.P.1

PROBLEMS FOR CHAPTER 4

(More problems involving higher order congruences may be found after Chapter 5.)

1. a) Show that if r belongs to the exponent δ (mod p) and $(\alpha, \delta) = \delta_1$ then

   r^α belongs to the exponent δ/δ_1 .

   b) Let g be a primitive root (mod p). For which α will g^α also be a

   primitive root?

2. Show that g is a primitive root mod p if and only if for each prime q

   dividing p-1, $g^{\frac{p-1}{q}} \not\equiv 1$ (mod p).

3. The usual method of finding primitive roots is by trial of primes 2, 3,

   5, 7, 11, 13,... until success is achieved. There is also some

   computational advantage in using 10 or -10 if they happen to be primitive

   roots. Find primitive roots for p = 17, 29, 41, 101 and 257. For which

   of these is 10 a primitive root? If 10 fails try -10.

4. Make an index table and anti-index table for p = 29. Label the index

   table NUM→INDEX and the anti-index INDEX→NUM.

5. Using the table of problem 4, solve those of the following equations

   which are solvable. Find all the roots. REMEMBER THAT INDEX

   CALCULATIONS ARE MOD p-1.

   a) $x^3 \equiv 22$ (mod 29)   e) $x^6 \equiv 13$ (mod 29)

   b) $x^4 \equiv 25$ (mod 29)   f) $x^8 \equiv 20$ (mod 29)

   c) $x^7 \equiv 14$ (mod 29)   g) $x^9 \equiv 11$ (mod 29)

   d) $x^7 \equiv 1$ (mod 29)   h) $x^{12} \equiv 14$ (mod 29).

4.P.2

6. Solve the following quadratic equations by dividing out the leading term
    and completing the square.

    a) $x^2+19x+19 \equiv 0 \pmod{29}$

    b) $3x^2+11x+19 \equiv 0 \pmod{29}$.

Problems 7-18 prove the existence of a primitive root mod p^e for p an odd
prime and $e > 1$. In general, g is a primitive root mod m if and only if, for
any a, if $(a,m) = 1$ then there is an α such that $a \equiv g^\alpha \pmod{m}$.

7. Let g_1 be a primitive root mod p, p a prime and let $g_1^{p-1} \equiv 1 \pmod{p^2}$.
    Show that $g = g_1+p$ is a primitive root mod p satisfying
    $g^{p-1} \not\equiv 1 \pmod{p^2}$. Thus a primitive root g exists satisfying
    $g^{p-1} \not\equiv 1 \pmod{p^2}$.

8. Show that if p is prime then $p | \binom{p}{k} = \dfrac{p!}{k!(p-k)!}$ for $1 < k < p-1$.

9. Show that for prime p and $k > 1$,
    If $a \equiv b \pmod{p^k}$ then $a^p \equiv b^p \pmod{p^{k+1}}$
    Hint: write $a = b+cp^k$.

10. Prove by induction on k that, for an odd prime p, integer r and $k > 2$
    $$(1+rp)^{p^{k-2}} \equiv 1+rp^{k-1} \pmod{p^k}.$$

11. Using the g of problem 7, show we may write $g^{p-1} = 1+rp$ where $p \nmid r$. Use
    the generalized theorem of Fermat to show $(g^{p-1})^{p^{e-1}} \equiv 1 \pmod{p^e}$ and
    Problem 10 to show $(g^{p-1})^{p^{e-2}} \not\equiv 1 \pmod{p^e}$. Thus show g^{p-1} belongs
    to the exponent p^{e-1} mod p^e.

12. Let the g of problem 7 belong to the exponent m mod p^e. Show using
    problem 11 that $p^{e-1} | m$. Set $m = p^{e-1}n$. Show $g^{p^{e-1}}$ is a primitive
    root mod p so that it belongs to the exponent $p-1$. Show $p-1 | n$. Recall
    (§19) that $m | \phi(p^e)$ and conclude that g belongs to the exponent $\phi(p^e)$, and
    thus the powers of g give all the relatively prime residues mod p^e.

4.P.3

13. Find a primitive root mod 5^2 and construct an index and anti-index table mod 25. Use it to solve $x^2 \equiv 6 \pmod{25}$ and $x^3 \equiv 18 \pmod{25}$.

14. Let g be a primitive root mod p^e. If g is odd show it is a primitive root mod $2p^e$. If g is even show $g+p^e$ is a primitive root mod $2p^e$.

15. Show that if m has a primitive root then the equation $x^2 \equiv 1 \pmod{m}$ has at most two solutions. Hint: if x is a solution then $x \equiv g^\alpha \pmod{m}$ with

    $0 \leqslant \alpha \leqslant m-2$.

16. Show that if p and q are distinct odd primes dividing m then $x^2 \equiv 1 \pmod{m}$ has four or more solutions.

17. Show that $x^2 \equiv 1 \pmod{2^e}$ has four solutions if $e \geqslant 3$. Hint: Consider $1+2^{e-1}$.

18. Using Problems 12, 14-17, show that the only values of m for which primitive roots exist are 2, 4, p^e, $2p^e$ for odd primes p and $e \geqslant 1$.

19. Use the results of §30 to answer the following:

    a) Is 58 an 8th power residue mod 61?

    b) Is 14 an 8th power residue mod 61?

    c) What is the relationship between being an 8th power residue and being a 4th power residue mod 61?

    d) How many eighth power residues are there mod 61?

20. Predict the lengths of the periods of the decimal expansions of the following fractions using the results of §31.

    a) $\dfrac{3}{7}$ b) $\dfrac{5}{11}$ c) $\dfrac{19}{31}$ d) $\dfrac{37}{101}$.

21. Show that if $(n,10) = 1$ and $(z,n) = 1$ then the length of the period of z/n must be a divisor of $\phi(n-1)$.

Chapter 5
Quadratic Residues

32. QUADRATIC RESIDUES MODULO m

DEFINITION Let m be a natural number and a an integer. Then a will be

called a quadratic residue or a quadratic non-residue mod m if and only

if $x^2 \equiv a$ (mod m) is solvable or not solvable.

This suggests the following two problems.

1st PROBLEM We are given a natural number m and seek to determine all the

quadratic residues mod m and the number of solutions of $x^2 \equiv a$ (mod m)

when a is a quadratic residue.

2nd PROBLEM We are given an integer a and seek to determine all the m

which have a as a quadratic residue, in other words, all the m for

which $x^2 \equiv a$ (mod m) is solvable.

33. CRITERION OF EULER AND THE LEGENDRE SYMBOL

The residue class containing 0 modulo m is formed of quadratic residues

for any m. Because of this, we will concern ourselves with quadratic

residues $a \not\equiv 0$ (mod m).

We will begin with the case m = p = (odd prime number). We could

have recourse to Theorem 1 of §30, but prefer to address ourselves

directly to stating and proving the following proposition.

5.33.1

<u>THEOREM</u> IF p is an odd prime and $(a,p) = 1$

$$x^2 \equiv a \pmod{p} \begin{cases} \text{is solvable if } a^{\frac{p-1}{2}} \equiv 1 \pmod{p}. \\ \text{is unsolvable if } a^{\frac{p-1}{2}} \not\equiv 1 \pmod{p}. \end{cases}$$

In other words, $x^2 \equiv a \pmod{p}$ is solvable if and only

if $a^{\frac{p-1}{2}} \equiv 1 \pmod{p}$.

<u>PROOF</u> OF NECESSITY: Let $x^2 \equiv a \pmod{p}$ with $a \not\equiv 0$ be solvable, and x_1 a

solution (which implies $(x_1,p) = 1$). Then $x_1^2 \equiv a \pmod{p}$ so

$$a^{\frac{p-1}{2}} \equiv (x_1^2)^{\frac{p-1}{2}} \equiv x_1^{p-1} \equiv 1 \pmod{p}$$

by the theorem of Fermat (§17).

<u>PROOF</u> OF SUFFICIENCY: Let $a^{\frac{p-1}{2}} \equiv 1 \pmod{p}$, whence $(a,p) = 1$. Then we

will have

$$\frac{p-1}{2} \text{ Ind } a \equiv \text{Ind } 1 \equiv p-1 \equiv 0 \pmod{p-1}$$

so

$$\frac{p-1}{2} \text{ Ind } a = k(p-1) \text{ with integer } k.$$

Consequently,

$$\text{Ind } a = 2k \text{ with integer } k.$$

5.33.2

Thus Ind $a \equiv 0 \pmod 2$ and it follows that the linear congruence

$$2 \cdot \text{Ind } x \equiv \text{Ind } a \pmod{p-1}$$

has a solution. This implies the solvability of $x^2 \equiv a \pmod p$.

We note that, since $(a,p) = 1$, we have $a^{p-1} - 1 \equiv 0 \pmod p$. However
$a^{p-1} - 1 \equiv (a^{(p-1)/2} - 1)(a^{(p-1)/2} + 1)$ and thus, if
$a^{(p-1)/2} - 1 \not\equiv 0 \pmod p$ we must have $a^{(p-1)/2} + 1 \equiv 0 \pmod p$.
Conversely, if $a^{(p-1)/2} + 1 \equiv 0 \pmod p$ then $a^{(p-1)/2} - 1 \not\equiv 0 \pmod p$. Thus
we arrive at the following

CRITERION OF EULER:

An $a \not\equiv 0 \pmod p$ is a quadratic residue or quadratic non-residue according
as

$$a^{\frac{p-1}{2}} \equiv 1 \pmod p \qquad \text{or} \qquad a^{\frac{p-1}{2}} \equiv -1 \pmod p.$$

Naturally, to find the quadratic residues relatively prime to p it
suffices to form the squares of the integers $1,2,\ldots,p-1$ and, indeed,
merely of $1,2,\ldots,(p-1)/2$, because if x_1 is one of $1,2,\ldots,(p-1)/2$
then $(p-x_1)^2 \equiv (-x_1)^2 \equiv x_1^2 \pmod p$.

For example, for $p=13$, we have the following quadratic residues
$$1^2 \equiv 1, \ 2^2 \equiv 4, \ 3^2 \equiv 9, \ 4^2 \equiv 3, \ 5^2 \equiv 12, \ 6^2 \equiv 10 \pmod{13}.$$
More generally, the quadratic residues of mod p, $p \geqslant 5$, are
$$1^2, 2^2, \ldots, \left(\frac{p-1}{2}\right)^2.$$
These are all incongruent mod p since if $1 \leqslant r < s \leqslant (p-1)/2$
then $s^2 - r^2 = (s-r)(s+r) \not\equiv 0 \pmod p$ because $1 \leqslant s-r \leqslant (p-1)/2$ and
$3 \leqslant s+r \leqslant p-1$.

5.33.3

<u>DEFINITION OF THE LEGENDRE SYMBOL</u> $\left(\frac{a}{p}\right)$.

The symbol $\left(\frac{a}{p}\right)$ is defined for $a \not\equiv 0 \pmod p$ and p an odd prime as follows

$$\left(\frac{a}{p}\right) = \begin{cases} 1 & \text{when } a \text{ is a quadratic residue mod } p \\ -1 & \text{when } a \text{ is a quadratic non-residue mod } p. \end{cases}$$

The symbol $\left(\frac{a}{p}\right)$ is pronounced "a over p".

Properties of $\left(\frac{a}{p}\right)$ are

    I) $a \equiv a' \pmod p \Rightarrow \left(\frac{a}{p}\right) = \left(\frac{a'}{p}\right)$.

    II) $\left(\frac{a}{p}\right) \equiv a^{(p-1)/2} \pmod p$. Specifically $\left(\frac{-1}{p}\right) = (-1)^{\frac{p-1}{2}}$.

    III) $\left(\frac{a \cdot b}{p}\right) \equiv \left(\frac{a}{p}\right)\left(\frac{b}{p}\right)$.

<u>PROOF</u> OF III. From II) follows

$$\left(\frac{a}{p}\right) \equiv a^{\frac{p-1}{2}} \pmod p \quad \text{and} \quad \left(\frac{b}{p}\right) \equiv b^{\frac{p-1}{2}} \pmod p$$

so

$$\left(\frac{a}{p}\right)\left(\frac{b}{p}\right) \equiv a^{\frac{p-1}{2}} b^{\frac{p-1}{2}} \equiv (ab)^{\frac{p-1}{2}} \equiv \left(\frac{a \cdot b}{p}\right) \pmod p.$$

However, $\left(\frac{a}{p}\right)\left(\frac{b}{p}\right) = \pm 1$ and $\left(\frac{a \cdot b}{p}\right) = \pm 1$ so, since $p \geqslant 3$, the congruence $\left(\frac{a}{p}\right)\left(\frac{b}{p}\right) \equiv \left(\frac{a \cdot b}{p}\right) \pmod p$ implies the equality $\left(\frac{a}{p}\right)\left(\frac{b}{p}\right) = \left(\frac{a \cdot b}{p}\right)$.

34.   STUDY OF THE CONGRUENCE $x^2 \equiv a \pmod p$

We will now study the congruence

$$x^2 \equiv a \pmod{p^r} \tag{34.1}$$

where $p > 2$ is prime and $(a,p) = 1$ and $r > 1$.

5.34.1

$\underline{\text{PROPOSITION}}$ For $x^2 \equiv a \pmod{p^r}$ to be solvable it is necessary and sufficient that $\left(\frac{a}{p}\right) = 1$ or, equivalently, that $x^2 \equiv a \pmod{p}$ be solvable.

$\underline{\text{Proof}}$ 1) The condition is necessary. Indeed, if $x_1^2 \equiv a \pmod{p^r}$ with $r \geq 2$ then also $x_1^2 \equiv a \pmod{p}$.

    2) The condition is sufficient. Let x_0 be a solution of
$$x_0^2 \equiv a \pmod{p^r} \text{ with } (x_0,p) = 1.$$
We will show that then $x^2 \equiv a \pmod{p^{r+1}}$ is solvable. For this purpose we set
$$x = x_0 + p^r y$$
and study the congruence $(x_0 + p^r y)^2 \equiv a \pmod{p^{r+1}}$ with unknown integer y. We have the congruence
$$x_0^2 + 2p^r x_0 y + p^{2r} y^2 \equiv a \pmod{p^{r+1}},$$
or, since $2r \geq r+1$
$$x_0^2 + 2p^r x_0 y \equiv a \pmod{p^{r+1}}.$$
However, by the hypothesis, $a - x_0^2 \equiv 0 \pmod{p^r}$ so $(a-x_0^2)/p^r$ is an integer. Accordingly the last congruence is equivalent to
$$2x_0 y \equiv \frac{a-x_0^2}{p^r} \pmod{p}.$$

This linear congruence for y is solvable because the coefficient $2x_0$ is relatively prime to p, which follows from $(x_0,p) = 1$ and $p > 2$ being prime. If y_0 is a solution then
$$x = x_0 + p^r(y_0 + kp) = x_0 + p^r y_0 + kp^{r+1}$$
is a solution of $x^2 \equiv a \pmod{p^{r+1}}$, which consequently is solvable.

    Now according to the hypothesis, $\left(\frac{a}{p}\right) = 1$ so $x^2 \equiv a \pmod{p}$ is solvable. By what we just proved, we know $x^2 \equiv a \pmod{p^2}$ is solvable,

5.34.2

then $x^2 \equiv a \pmod{p^3}$ is solvable and, generally, $x^2 \equiv a \pmod{p^r}$ is solvable for $r > 1$.[*]

Since, according to property I of the symbol $\left(\frac{a}{p}\right)$, the equality $\left(\frac{a}{p}\right) = 1$ implies $\left(\frac{a+hp}{p}\right) = 1$ for $h \in \mathbb{Z}$, if a (with $(a,p) = 1$) is a quadratic residue mod p^r then each $a+hp$ (with $h \in \mathbb{Z}$) is also a quadratic residue mod p^r. How many of these are incongruent mod p^r? We have

$$a+h_1 p \equiv a+h_2 p \pmod{p^r} \Rightarrow h_1 - h_2 \equiv 0 \pmod{p^{r-1}}$$

and conversely $h_1 - h_2 \equiv 0 \pmod{p^{r-1}} \Rightarrow a+h_1 p \equiv a+h_2 p \pmod{p^r}$. Consequently to a definite quadratic residue a mod p correspond p^{r-1} quadratic residues mod p^r which are incongruent to each other.

Moreover, to two incongruent quadratic residues a_1 and a_2 mod p correspond incongruent $a_1+h_1 p$ and $a_2+h_2 p$ $(\bmod\ p_r)$:

$$a_1 \not\equiv a_2 \bmod p \Rightarrow a_1 + h_1 p \not\equiv a_2 + h_2 p \pmod{p^r}.$$

Hence, since their are $(p-1)/2$ incongruent quadratic residues mod p, the number of quadratic residues mod p^r is

$$\frac{p-1}{2} \cdot p^{r-1} = \frac{\phi(p^r)}{2} \quad \text{(where } \phi \text{ is Euler's function)}.$$

Finally we examine how many incongruent mod p^r solutions there are to $x^2 \equiv a \pmod{p^r}$ when $\left(\frac{a}{p}\right) = 1$, that is, when $x^2 \equiv a \pmod{p^r}$ has at least one solution x_0. The congruence also has the solution $-x_0$, which is incongruent to x_0 mod p^r since $2x_0 \not\equiv 0 \bmod p^r$. However, any solution x_1 of

---

[*]In the problems will be found a computationally more efficient procedure.

5.34.3

$x^2 \equiv a \pmod{p^r}$ satisfies the relation

$$(x_1 - x_0)(x_1 + x_0) \equiv (x_1^2 - x_0^2) \equiv 0 \pmod{p^r}$$

so we have

$$\text{either } x_1 \equiv x_0 \pmod{p^r} \text{ or } x_1 \equiv -x_0 \pmod{p^r}$$

for, in the contrary case, p would divide both $x_1 - x_0$ and $x_1 + x_0$ and hence their difference $2x_0$, which is false since $(x_0, p) = 1$. Hence $x^2 \equiv a \pmod{p^r}$ has just two incongruent solutions when $\left(\frac{a}{p}\right) = 1$.

<u>EXAMPLE</u> Consider the congruence $x^2 \equiv 2 \pmod{7^3}$. We have $\left(\frac{2}{7}\right) \equiv 2^{(7-1)/2} = 2^3 = 8 \equiv 1 \pmod{7}$ so $\left(\frac{2}{7}\right) = 1$. Consequently the congruence is solvable. In order to find a solution we take an x_0 with $x_0^2 \equiv 2 \pmod{7}$, for example $x_0 = 3$, and set $x = 3 + 7y$. We determine y such that $x^2 = 9 + 2 \cdot 21y + 49y^2 \equiv 2 \pmod{7^2}$, so $42y \equiv -7 \pmod{7^2}$ and thus $6y \equiv -1 \pmod{7}$, from which we get $y \equiv 1 \pmod{7}$. Consequently $x = 3 + 7 \cdot 1 = 10$ is a solution of $x^2 \equiv 2 \pmod{7^2}$. We continue, setting $x = 10 + 7^2 y$ and solving for y the congruence

$$x^2 \equiv 100 + 2 \cdot 49 \cdot 10y + 7^4 y^2 \equiv 2 \pmod{7^3}$$

or the equivalent

$$20 \cdot 49 y \equiv -98 \pmod{7^3}$$

which is equivalent to

$$20y \equiv -2 \pmod{7}.$$

Consequently $y \equiv 2 \pmod{7}$ and $x = 10 + 2 \cdot 49 = 108$ is a solution of $x^2 \equiv 2 \pmod{7^3}$. Verification: $108^2 = 11664$ and $11662 = 34 \cdot 7^3$. So the solutions of $x^2 \equiv 2 \pmod{7^3}$ are $x \equiv \pm 108 \pmod{7^3}$.

5.35.1

35. STUDY OF THE CONGRUENCE $x^2 \equiv a \pmod{2^k}$

We now have to study the congruence $x^2 \equiv a \pmod{2^k}$ where k is a natural

number. We distinguish four cases:

   I) $x^2 \equiv a \pmod 2$. We easily determine that

       $x^2 \equiv 0 \pmod 2$ has the solution $x \equiv 0 \pmod 2$ and

       $x^2 \equiv 1 \pmod 2$ has the solution $x \equiv 1 \pmod 2$.

   Thus the quadratic residues mod 2 are 0 and 1.

   II) $x^2 \equiv a \pmod{2^2}$. We note that

       $x^2 \equiv 0 \pmod 4$ has two solutions $x \equiv 0,2 \pmod 4$ and

       $x^2 \equiv 1 \pmod 4$ has two solutions $x \equiv 1,3 \pmod 4$ and thus

       $x^2 \equiv 2$ and $x^2 \equiv 3 \pmod 4$ have no solutions.

   Thus the quadratic residues mod 4 are 0,1.

   III) $x^2 \equiv a \pmod{2^3}$. The even quadratic residues mod 8 are 0,4;

2 and 6 are non-residues. We easily find

       $x^2 \equiv 0 \pmod 8$ has the solutions $x = 4k$ with $k \in \mathbb{Z}$ and

       $x^2 \equiv 4 \pmod 8$ has the solutions $x = 4k+2$ with $k \in \mathbb{Z}$.

   Let now a be an odd integer. Then any solution of

       $x^2 \equiv a \pmod 8$ will be an odd integer: $x = 2n+1$ and thus

       $x \equiv \pm1, \pm3 \pmod 8$. Since

$$(\pm1)^2 \equiv 1 \text{ and } (\pm3)^2 \equiv 1 \pmod 8.$$

The congruence $x^2 \equiv a \pmod 8$ is solvable if and only if

$a \equiv 1 \pmod 8$. Thus the odd quadratic residues $\pmod 8$ are the

5.35.2

elements of the residue class containing 1 (mod 8) and only these. Moreover, the number of incongruent mod 8 solutions of $x^2 \equiv 1$ (mod 8) is equal to 4, since the solutions of $x^2 \equiv 1$ (mod 8) are 1, 3, 5, 7.

IV) $x^2 \equiv a$ (mod 2^{k+1}) with $k \geqslant 2$ and $a \neq 0$. We will investigate the odd quadratic residues of 2^{k+1}, because the determination of the even quadratic residues $a \neq 0$ can be reduced to the determination of the odd quadratic residues.

<u>PROPOSITION</u> The congruence $x^2 \equiv a$ (mod 2^{k+1}) with $k \geqslant 2$ and odd a is solvable, if and only if $a \equiv 1$ (mod 8).

<u>Proof</u> By III, the assertion is true for $k = 2$. Supposing it true for $k \geqslant 2$, let $x_0^2 \equiv a$ (mod 2^{k+1}). Then $x_0^2 \equiv a$ (mod 8) and, according to III, $a \equiv 1$ (mod 8) so the condition $a \equiv 1$ (mod 8) is necessary. I say it is also sufficient. Actually, recall that x_0 and a are odd and set
$$x = x_0 + 2^k y.$$
We seek to determine y so that x satisfies
$$x^2 = x_0^2 + 2^{k+1} x_0 y + 2^{2k} y^2 \equiv a \pmod{2^{k+2}}.$$
Since $2k \geqslant k+2$, this congruence for y can be simplified to
$$x_0^2 + 2^{k+1} x_0 y \equiv a \pmod{2^{k+2}}$$
and so
$$2^{k+1} x_0 y \equiv a - x_0^2 \pmod{2^{k+2}}.$$
Since, by hypothesis, $x^2 \equiv a$ (mod 2^{k+1}), this last is equivalent to
$$x_0 y \equiv \frac{a - x_0^2}{2^{k+1}} \pmod{2}.$$

5.35.3

In this last linear congruence for y, the coefficient x_0 is relatively prime to 2, so the congruence is solvable, and thus the congruence $x^2 \equiv a \pmod{2^{k+2}}$ is also solvable.

We have shown that if the proposition is true for $k \geqslant 2$ it is also true for $k+1$. Thus, by mathematical induction, it is true for all $k \geqslant 2$.

Next we note that the number of incongruent relatively prime to 2 quadratic residues mod 2^k is identical to the number of incongruent relatively prime residues mod 2^k which are congruent to 1 mod 8. There are thus $\phi(2^k)/4 = 2^{k-3}$ incongruent relatively prime to 2 quadratic residues mod 2^k.

Let us now investigate the number of incongruent solutions of $x^2 \equiv a \pmod{2^k}$ with $a \equiv 1 \pmod 8$ (and so odd) and $k \geqslant 4$. Let x_0 be a solution: $x_0^2 \equiv a \pmod{2^k}$. Note x_0 is odd. Let x be any other solution. Then x is odd and

$$x^2 \equiv x_0^2 \pmod{2^k}$$

so

$$(x-x_0)(x+x_0) \equiv 0 \pmod{2^k}.$$

Since x and x_0 are odd, $x-x_0$ and $x+x_0$ are even, so we have

$$\frac{x-x_0}{2} \cdot \frac{x+x_0}{2} \equiv 0 \pmod{2^{k-2}}.$$

The factors $(x+x_0)/2$ and $(x-x_0)/2$ cannot both be even, for then their difference x_0 would be even. For the same reason they cannot both be odd. Thus we will have

5.35.4

either $\dfrac{x-x_0}{2}$ odd and thus $\dfrac{x+x_0}{2} \equiv 0 \pmod{2^{k-2}}$

or $\dfrac{x+x_0}{2}$ odd and thus $\dfrac{x-x_0}{2} \equiv 0 \pmod{2^{k-2}}$.

So we will have

$$\text{either } x \equiv -x_0 \pmod{2^{k-1}} \text{ or } x \equiv x_0 \pmod{2^{k-1}}. \qquad (35.1)$$

Conversely, if one of these last is true then we will have not only

$$(x-x_0)(x+x_0) \equiv 0 \pmod{2^{k-1}}$$

but also

$$(x-x_0)(x+x_0) \equiv 0 \pmod{2^k}$$

because one of the factors $x-x_0$ and $x+x_0$ is congruent to 0 mod 2^{k-1} and the other is even since x and x_0 are both odd, x_0 by hypothesis and x by (35.1). Consequently

$$x^2 - x_0^2 \equiv 0 \pmod{2^k}$$

and thus (35.1) implies that $x^2 \equiv x_0^2 \equiv a \pmod{2^k}$.

Hence all the solutions of $x^2 \equiv a \pmod{2^k}$ are determined by (35.1) and thus are the integers

$$x = x_0 + t_1 2^{k-1} \text{ and } x = -x_0 + t_2 2^{k-1} \text{ with integer } t_1, t_2.$$

The incongruent mod 2^k solutions are thus*

$$x_0, \ x_0 + 2^{k-1}, \ -x_0, \ -x_0 + 2^{k-1}.$$

<u>EXAMPLE</u> $x^2 \equiv 9 \pmod{16}$. Let $x_0 = 3$. Then the four solutions are 3, $3+8 \equiv 11$, $-3 \equiv 13$, $-3+8 \equiv 5$.

*These are easily shown to be incongruent mod 2^k; see problems.

5.36.1

36. STUDY OF THE CONGRUENCE $x^2 \equiv a \bmod m$ WITH $(a,m) = 1$

LEMMA Let $f(x)$ be a polynomial with integer coefficients, and $m = m_1 \ldots m_k$ with $(m_i, m_j) = 1$ for $i \neq j$. Then

$$f(x) \equiv 0 \pmod{m} \text{ is solvable}$$

if and only if the system

$$f(x) \equiv 0 \pmod{m_1}, \ f(x) \equiv 0 \pmod{m_2}, \ldots, f(x) \equiv 0 \pmod{m_k}$$

is solvable.

Proof Since $f(x) \equiv 0 \pmod{m} \Rightarrow f(x) \equiv 0 \pmod{m_j}$ $(j=1,\ldots,k)$, the sufficiency is obvious. For the necessity let $f(x) \equiv 0 \pmod{m_j}$ have $\lambda_j \geq 1$ incongruent solutions mod m_j for $j=1,\ldots,k$. Let $r_1,\ldots,r_k$ be some system of solutions: $f(r_j) \equiv 0 \pmod{m_j}$, $j = 1,\ldots,k$. According to the results of §22, we may determine an integer x_0 for which

$$x_0 \equiv r_j \pmod{m_j} \text{ for } j = 1,\ldots,k.$$

Then we will have

$$f(x_0) \equiv 0 \pmod{m_j} \text{ for } j = 1,\ldots,k$$

and thus, since $(m_i, m_j) = 1$ for $i \neq j$

$$f(x_0) \equiv 0 \pmod{m}.$$

The number of solutions is $\lambda_1 \lambda_2 \ldots \lambda_k$, since two distinct systems $(r_1,\ldots,r_k)$ and $(r_1',\ldots,r_k')$ determine two incongruent solutions x_0 and x_0' mod $m_1 \ldots m_k = m$.

Now let $m = 2^\alpha p_1^{\alpha_1} \ldots p_k^{\alpha_k}$ with $\alpha \geq 0$, $\alpha_j \geq 1$, $j = 1,\ldots,k$. We consider the congruence

$$x^2 \equiv a \pmod{m} \text{ with } (a,m) = 1. \tag{36.1}$$

5.36.2

Then $(a,p_j) = 1$ and $(a,2) = 1$ if $\alpha > 0$.

I. Let $\alpha = 0$. For (36.1) to be solvable it is necessary and sufficient that

$$\left(\frac{a}{p_1}\right) = \left(\frac{a}{p_2}\right) = \ldots = \left(\frac{a}{p_k}\right) = 1.$$

The number of incongruent quadratic residues mod $m = p_1^{\alpha_1} \ldots p_k^{\alpha_k}$ is

$$p_1^{\alpha_1 - 1} \frac{p_1 - 1}{2} \ldots p_k^{\alpha_k - 1} \frac{p_k - 1}{2} = \frac{\phi(m)}{2^k}$$

since $p_j^{\alpha_j - 1}(p_j - 1)/2$ is the number of relatively prime quadratic residues mod $p_j^{\alpha_j}$, $j = 1,\ldots,k$. The number of incongruent solutions of $x^2 \equiv a \pmod{m}$, where a is a relatively prime to m quadratic residue, is 2^k since each $x^2 \equiv a \pmod{p_j^{\alpha_j}}$ has two incongruent solutions mod $p_j^{\alpha_j}$ for $j = 1,\ldots,k$.

EXAMPLE: $x^2 \equiv a \pmod{15}$ has solutions for $a \equiv 1, 4 \pmod{15}$ since $\left(\frac{1}{3}\right) = \left(\frac{1}{5}\right) = 1$ and $\left(\frac{4}{3}\right) = \left(\frac{4}{5}\right) = 1$. Since their are $\phi(15)/2^2 = [3(1-\frac{1}{3}) \cdot 5(1-\frac{1}{5})] \div 2^2 = 2$ relatively prime to 15 quadratic residues, the other relatively prime residues 2, 7, 8, 11, 13, 14 are quadratic non-residues mod 15. The number of solutions of $x^2 \equiv 1$ and $x^2 \equiv 4 \pmod{15}$ is four for each, as is easily verified.

5.36.3

II. Let $m = 2p_1^{\alpha_1}\ldots p_k^{\alpha_k}$. The congruence $x^2 \equiv a \pmod{m}$ with $(a,m) = 1$ has solutions if and only if the congruences
$$x^2 \equiv a \pmod{2} \text{ and } x^2 \equiv a \pmod{p_j^{\alpha_j}} \quad j = 1,\ldots,k$$
have solutions. But this is equivalent to
$$\left(\frac{a}{p_1}\right) = \left(\frac{a}{p_2}\right) = \ldots = \left(\frac{a}{p_k}\right) = 1$$
because $x^2 \equiv a \pmod{2}$ is solvable for any a. The number of incongruent to m relatively prime quadratic residues of $m = 2p_1^{\alpha_1}\ldots p_k^{\alpha_k}$ is again $\phi(m)/2^k$ and the number of solutions of a solvable $x^2 \equiv a \pmod{m}$, $(a,m) = 1$, is again 2^k.

EXAMPLE $x^2 \equiv a \pmod{18}$. The number of relatively prime to 18 quadratic residues is $\phi(18)/2^1 = 3$. They are 1, 7, 13. The number of solutions for each is $2^1 = 2$.

III. $m = 2^2 p_1^{\alpha_1}\ldots p_k^{\alpha_k}$. The congruence $x^2 \equiv a \pmod{m}$ with $(a,m) = 1$ has solutions if and only if
$$a \equiv 1 \pmod{4} \text{ and } \left(\frac{a}{p_1}\right) = \left(\frac{a}{p_2}\right) = \ldots = \left(\frac{a}{p_k}\right) = 1.$$

The number of relatively prime to m quadratic residues is

$$\phi(m)/2^{k+1} = \phi(p_1^{\alpha_1}\ldots p_k^{\alpha_k})/2^k.$$ The number of solutions will be 2^{k+1}.

5.36.4

EXAMPLE $x^2 \equiv a \pmod{20}$. The number of quadratic residues is $\phi(5^1)/2^1 = 2$. They are 1, 9. The number of solutions for each congruence will be 4. For example $x^2 \equiv 9 \pmod{20}$ has solutions

$x \equiv 3,\ 7,\ 13,\ 17$.

IV. $m = 2^\alpha p_1^{\alpha_1} \ldots p_k^{\alpha_k}$ with $\alpha \geqslant 3$. The congruence $x^2 \equiv a \pmod{m}$ with $(a,m) = 1$ has solutions if and only if

$$a \equiv 1 \pmod{8},\ \left(\frac{a}{p_1}\right) = \left(\frac{a}{p_2}\right) = \ldots = \left(\frac{a}{p_k}\right) = 1.$$

The number of relatively prime to m quadratic residues is

$$(\phi(2^\alpha)/4)(\phi(p_1^{\alpha_1})/2)\ldots(\phi(p_k^{\alpha_k})/2) = 2^{\alpha-3} \cdot \phi(p_1^{\alpha_1} \ldots p_k^{\alpha_k})/2^k =$$
$$2^{\alpha-k-3} \cdot \phi(p_1^{\alpha_1} \ldots p_k^{\alpha_k}).$$ The number of solutions will be 2^{k+2}.

EXAMPLE $x^2 \equiv a \pmod{288 = 2^5 \cdot 3^2}$. The number of relatively prime quadratic residues is $2^{5-1-3}\phi(3^2) = 2^1 \cdot 3^2(1-\frac{1}{3}) = 12$ They are 1, 25, 49, 121, 169, 73, 241, 265, 97, 217, 193, 145. The number of solutions for $x^2 \equiv a \pmod{2^5 \cdot 3^2}$ for a equaling one of these values is $2^{1+2} = 2^3 = 8$. For example $x^2 \equiv 1 \pmod{2^5 \cdot 3^2}$ has solutions 1, 17, 127, 143, 145, 161, 271, 287.

37. GENERALIZATION OF THE THEOREM OF WILSON

Let m > 2 be a natural number, a an integer with $(a,m) = 1$, and $x^2 \equiv a \pmod{m}$ a solvable congruence. If x_1 is a solution, then $-x_1$ is also a solution and

5.37.1

indeed $-x_1 \not\equiv x_1 \pmod{m}$ because $-x_1 \equiv x_1 \pmod{m} \Rightarrow 2x_1 \equiv 0 \pmod{m} \Rightarrow$ $(2x_1,m) = m > 2$. But $(x_1,m) = (a,m) = 1$ so $(2x_1,m) = 1$ or $(2x_1,m) = 2$.

The number λ of incongruent solutions of $x^2 \equiv a \pmod{m}$ is even, as we saw in the previous section and which, moreover, we could prove with mathematical induction. Let us represent the solutions as

$$x_1, \; x_2, \; \dots \; , \; x_{\lambda/2}, \; -x_1, \; -x_2, \; \dots \; , -x_{\lambda/2}.$$

Then we will have

$$\prod_{j=1}^{\lambda/2} x_j(-x_j) = (-1)^{\lambda/2} \prod_{j=1}^{\lambda/2} x_j^2 \equiv (-1)^{\lambda/2} a^\lambda \pmod{m}.$$

Let now $r_1, \; \dots \; , \; r_{\phi(m)}$ be a reduced system of residues modulo m. The linear inequality

$$r_i x \equiv a \pmod{m} \quad (i = 1,\dots,\phi(m))$$

has as a solution a certain r_j from among $r_1, \; \dots \; , \; r_{\phi(m)}$; that is, for some j, $r_i r_j \equiv a \pmod{m}$. We say that r_i is associated with r_j if $r_i \neq r_j$ and r_i is associated with itself if $r_i = r_j$, (which can occur only if a is a quadratic residue mod m). We distinguish two cases:

1st CASE a is a quadratic non-residue mod m. Then there is no element associated with itself and each r_i is associated with a distinct r_j. Hence we have $\phi(m)/2$ pairs and thus

$$r_1 r_2 \cdots \; r_{\phi(m)} \equiv a^{\phi(m)/2} \pmod{m}.$$

2nd CASE a is a quadratic residue mod m. Let $x^2 \equiv a \pmod{m}$ have the solutions $x_1, \; -x_1, \; x_2, \; \dots \; , \; x_{\lambda/2}, \; -x_{\lambda/2}$. Each of these is associated

5.37.2

with itself. Now we note the following: In the system
$\{r_1, r_2, \ldots, r_{\phi(m)}\}$, λ integers are congruent mod m to corresponding
elements of $x_1, -x_1, \ldots, x_{\lambda/2}, -x_{\lambda/2}$; so their product is congruent to

$$\prod_{j=1}^{\lambda/2} x_j(-x_j) \equiv (-1)^{\lambda/2} a^{\lambda/2} \quad (\text{mod } m).$$

The remaining elements of the system $\{r_1, \ldots, r_{\phi(m)}\}$ fall into
$(\phi(m)-\lambda)/2$ pairs, each a pair of associated elements, so their product is
congruent to $a^{(\phi(m)-\lambda)/2}$. In consequence, in the second case
$$r_1 r_2 \cdots r_{\phi(m)} \equiv (-1)^{\lambda/2} a^{\lambda/2} a^{(\phi(m)-\lambda)/2} = (-1)^{\lambda/2} a^{\phi(m)/2}.$$
Thus we have modulo m

$$r_1 r_2 \cdots r_{\phi(m)} \equiv \begin{cases} a^{\phi(m)/2} & \text{when a is a quadratic non-residue mod m} \\ (-1)^{\lambda/2} a^{\phi(m)/2} & \text{when a is a quadratic residue mod m} \end{cases} \qquad (37.1)$$

where λ is the number of incongruent solutions of $x^2 \equiv a$ (mod m).

<u>REMARK</u> The number λ does not depend on a, but only on the prime

decomposition of m, as we saw in §36.

Specialize now to a = 1. We find ourselves in the 2nd case, and have
$$r_1 \cdots r_{\phi(m)} \equiv (-1)^{\lambda/2} \quad (\text{mod } m)$$
where λ is the number of solutions of $x^2 \equiv 1$ (mod m).

We will determine when $\lambda/2$ is even. Let $m = 2^\alpha p_1^{\alpha_1} \cdots p_k^{\alpha_k}$.

If $\alpha = 0, 1$ then $\lambda = 2^k$ and $\frac{\lambda}{2} = 2^{k-1}$; thus $\frac{\lambda}{2}$ is odd for k = 1.

If $\alpha = 2$ then $\lambda = 2^{k+1}$ and $\frac{\lambda}{2} = 2^k$; thus $\frac{\lambda}{2}$ is odd for k = 0.

5.37.3

If $\alpha \geqslant 3$ then $\lambda = 2^{k+2}$ and $\frac{\lambda}{2} = 2^{k+1}$; thus $\frac{\lambda}{2}$ is never odd.

Hence we may formulate the following generalization of the theorem of Wilson:

<u>THEOREM</u> We have, modulo m

$$r_1 r_2 \cdots r_{\phi(m)} \equiv \begin{cases} -1 \text{ when } m = p_1^{\alpha_1},\ 2p_1^{\alpha_1},\ 4 \\ 1 \text{ for any other } m. \end{cases}$$

<u>GENERALIZATION OF THE THEOREM OF FERMAT</u> Below we will exhibit another proof of the generalized theorem of Fermat

Since 1 is a quadratic residue mod m we find, for $a = 1$ and $m > 2$, that

$$r_1 \cdots r_{\phi(m)} \equiv (-1)^{\lambda_0/2} \pmod{m}$$

where λ_0 is the number of incongruent solutions of $x^2 \equiv 1 \pmod{m}$*.

We introduce this value of $r_1 \cdots r_{\phi(m)}$ into (37.1) which we now write

$$a^{\phi(m)/2} \equiv \begin{cases} r_1 \cdots r_{\phi(m)} \equiv (-1)^{\lambda_0/2} \text{ when } a \text{ is a quadratic non-residue mod } m \\ (-1)^{\lambda/2} r_1 \cdots r_{\phi(m)} \equiv (-1)^{\lambda/2}(-1)^{\lambda_0/2} \text{ when } a \text{ is a quadratic residue mod } m. \end{cases}$$

Squaring both sides, we now find, for $m > 2$ and any a with $(a,m) = 1$

$$a^{\phi(m)} \equiv 1 \pmod{m}.$$

This congruence is also obviously true for $m = 2, 1$.

* We know from §36 that $\lambda_0 = \lambda$, but we do not wish to use this fact in the proof.

38. TREATMENT OF THE SECOND PROBLEM OF §32

We will now treat the second problem of §32. We are given the integer $a \neq 0$. We seek to determine the natural numbers m for which $(a,m) = 1$ and $x^2 \equiv a \pmod{m}$ is solvable. We may set

$$m = 2^{\alpha} p_1^{\alpha_1} \ldots p_k^{\alpha_k} \quad \text{with } \alpha \geqslant 0 \text{ and } \alpha_j > 0 \text{ for } j = 1,\ldots,k.$$

For m to satisfy the above conditions, obviously the congruences $x^2 \equiv a \pmod{p_j}$ $j = 1,\ldots,k$ must be solvable, that is, we must have $\left(\frac{a}{p_j}\right) = 1$. In addition, for the case of even a ($a \equiv 0 \pmod 2$), m must be odd since $(a, m) = 1$, and in the case of odd a ($a \equiv 1 \pmod 2$):

if $a \equiv 1 \bmod 8$ then α may be any non-negative integer

if $a \not\equiv 1 \bmod 8$ but $a \equiv 1 \pmod 4$ then $\alpha = 0,1,2$

if $a \not\equiv 1 \bmod 4$ but $a \equiv 1 \pmod 2$ then $\alpha = 0,1$.

According to what was shown in §32-37, the above necessary conditions are also sufficient for m to have the required properties, that is, for a to be a relatively prime to m quadratic residue.

Thus, in the required m, the 2 may appear to any power 2^{α} with $\alpha > 0$, if $a \equiv 1$ (mod 8), but only as $2^0, 2^1, 2^2$ if $a \equiv 5$ (mod 8), and only as $2^0, 2^1$ if $a \equiv 3,7$ (mod 8). The 2 may not appear at all in m if the given a is even. Any prime p for which $\left(\frac{a}{p}\right) = 1$ may appear in m, but if $\left(\frac{a}{p}\right) = -1$ then p may not be a factor of m. Consequently the problem reduces to the determination of the odd primes p for which $(a, p) = 1$ and $\left(\frac{a}{p}\right) = 1$.

Since $\left(\frac{a}{p}\right)\left(\frac{a'}{p}\right) = \left(\frac{aa'}{p}\right)$ with $a, a' \not\equiv 0 \pmod p$, and since any integer a may be written $a = \pm\, 2^{\alpha}\, p_1^{\alpha_1}\ldots p_k^{\alpha_k}$, we have, for p satisfying $(a,p) = 1$,

$$\left(\frac{a}{p}\right) = \left(\frac{\pm\, 2^{\alpha} p_1 \ldots p_k^{\alpha_k}}{p}\right) = \left(\frac{\pm 1}{p}\right) \left(\frac{2}{p}\right)^{\alpha} \left(\frac{p_1}{p}\right)^{\alpha_1} \ldots \left(\frac{p_k}{p}\right)^{\alpha_k}.$$

Thus the problem reduces to the three cases

$$a = -1, \qquad a = 2, \qquad a = q, \text{ an odd prime with } q \neq p.$$

5.39.1

39. STUDY OF $(\frac{-1}{p})$ AND APPLICATIONS

We ask, for which $p>2$ is $(\frac{-1}{p}) = 1$?

As we know,

$$(\frac{-1}{p}) \equiv (-1)^{(p-1)/2} \pmod{p}.$$

This congruence implies the equality $(\frac{-1}{p}) = (-1)^{(p-1)/2}$ because

$$|(\frac{-1}{p}) - (-1)^{(p-1)/2}| < 2 \text{ and } p \geqslant 3.$$

From $(\frac{-1}{p}) = (-1)^{(p-1)/2}$ it follows that

$$(\frac{-1}{p}) = 1 \quad \text{for} \quad p = 4k + 1 \quad \text{with integer } k$$

and $\quad (\frac{-1}{p}) = -1 \quad \text{for} \quad p = 4k + 3.$

Thus we have the theorem

PROPOSITION 1. -1 is a quadratic residue mod $p = 4k + 1$ and a quadratic non-residue mod $p = 4k + 3$.

PROPOSITION 2. The odd prime divisors of an integer $x^2 + 1$ with $k \in \mathbb{Z}$ are of the form $4k + 1$.

PROOF From the hypothesis $p | x^2 + 1$ it follows that $x^2 \equiv -1 \pmod{p}$ is solvable, so -1 is a quadratic residue mod p and, according to proposition 1, p has the form $4k + 1$.

PROPOSTION 3. There are infinitely many primes of the form $4k + 1$ (k a natural number).

PROOF Let $p_1 \ldots, p_\lambda$ be primes of the form $4k + 1$. (For example $p_1 = 5$, $p_2 = 13, \ldots$). We form the integer

$$4 \, p_1^2 \, p_2^2 \ldots p_\lambda^2 + 1. \tag{39.1}$$

If this is prime, then we have proved that primes exist of the form $4k + 1$ distinct from $p_1, \ldots, p_\lambda$. If (39.1) is not prime then, being odd, it will have

5.39.2

an odd prime divisor and, according to proposition 2, this will be of the form 4k+1. However it will be distinct from $p_1,\ldots,p_\lambda$, since in the contrary case it would divide 1. Thus $p_1,\ldots,p_\lambda$ cannot exhaust the primes of the form 4k+1 and consequently there are infinitely many.

<u>PROPOSITION 4.</u> There exist infinitely many primes of the form 4k+3.

   <u>Proof</u> Let $q_1,\ldots,q_n$ be primes of the form 4k+3. (For example, $q_1=7$, $q_2=11,\ldots$). We form the integer

$$q_1^2\ q_2^2\ldots\ q_n^2 + 2. \tag{39.2}$$

   Since $q_j^2 \equiv 3^2 \equiv 1 \pmod 4$, this integer is of the form 4k+3. Hence not all its prime divisors can be of the form 4k+1, because the product of such factors would again be of the form 4k+1. Thus, there must exist at least one prime divisor of (39.2) of the form 4k+3. This must be distinct from $q_1,\ldots,q_n$ since in the contrary case it would divide 2. Hence $q_1,\ldots,q_n$ cannot exhaust the primes of the form 4k+3 and consequently there are infinitely many.

40. THE LEMMA OF GAUSS

   Let $p \geqslant 3$ be an odd prime and a an integer with $(a,p) = 1$. We will represent the integer $(p-1)/2$ by p'. The reduced system of residues with least absolute values is

$$1,2,\ldots,p',-1,-2,\ldots,-p'. \tag{40.1}$$

Among the numbers $1{\cdot}a,2{\cdot}a,\ldots,p'{\cdot}a$, some, when reduced mod p to their least absolute residue (40.1), have a negative residue and some have a positive residue. Let us designate the first by

5.40.1

$\nu_1 a \equiv -\alpha_1, \ldots, \nu_\mu a \equiv -\alpha_\mu$ (mod p) where $-\alpha_i$ are negative residues

and the second by

$\pi_1 a \equiv \beta_1, \ldots, \pi_{p'-\mu} a \equiv \beta_{p'-\mu}$ (mod p) where β_j are positive residues.

Along with the ν_i the α_i are incongruent mod p and the same with the β_j.

Each α_i is also incongruent with each β_j, since

$$\alpha_i \equiv \beta_j \text{ (mod p)} \Rightarrow -\nu_i a \equiv \pi_j a \Rightarrow \pi_j + \nu_i \equiv 0 \text{ (mod p)};$$

but $\pi_j + \nu_i \not\equiv 0$ (mod p) since $1 \leqslant \pi_j, \nu_i \leqslant p'$ so $2 \leqslant \pi_j + \nu_i \leqslant 2p' = p-1$.

Consequently the sets $\{\alpha_1, \ldots, \alpha_\mu, \beta_1, \ldots, \beta_{p'-\mu}\}$ and $\{\nu_1 \ldots \nu_\mu, \pi_1, \ldots, \pi_{p'-\mu}\}$

are each identical to $\{1, 2, \ldots, p'\}$. Hence we have the congruence

$$a^{p'}(1 \cdot 2 \ldots p') = a^{p'}(\nu_1 \ldots \nu_\mu \, \pi_1 \ldots \pi_{p'-\mu})$$

$$= (a\nu_1) \ldots (a\nu_\mu)(a\pi_1) \ldots (a\pi_{p'-\mu})$$

$$\equiv (-\alpha_1) \ldots (-\alpha_\mu)(\beta_1) \ldots (\beta_{p'-\mu}) \quad \text{(mod p)}$$

$$\equiv (-1)^\mu (\alpha_1 \ldots \alpha_\mu \beta_1 \ldots \beta_{p'-\mu}) \quad \text{(mod p)}$$

$$\equiv (-1)^\mu (1 \cdot 2 \ldots p') \quad \text{(mod p)}.$$

5.40.2

and thus

$$a^{(p-1)/2} = a^{p'} \equiv (-1)^{\mu} \pmod{p}.$$

However,

$$\left(\frac{a}{p}\right) \equiv a^{(p-1)/2} \pmod{p}$$

so we will have

$$\left(\frac{a}{p}\right) \equiv (-1)^{\mu} \pmod{p}$$

and thus

$$\left(\frac{a}{p}\right) = (-1)^{\mu}.$$

We have proved the following proposition, called the lemma of Gauss:

<u>LEMMA OF GAUSS</u>

$$a \begin{cases} \text{is a quadratic residue mod p when among the integers} \\ a, 2a, \ldots, p'a \text{ the number of them with negative least absolute} \\ \text{residues is even.} \\ \\ \text{is a quadratic non-residue when among the integers} \\ a, 2a, \ldots, p'a \text{ the number of them with negative least absolute} \\ \text{residues is odd.} \end{cases}$$

For example, 2 is a quadratic residue mod 7 since among the numbers

$$2 \cdot 1 \equiv 2 \qquad 2 \cdot 2 \equiv 4 \equiv -3 \qquad 2 \cdot 3 \equiv 6 \equiv -1$$

two have negative least absolute residues. But 3 is not a quadratic residue
since among the numbers

$$3 \cdot 1 \equiv 3 \qquad 3 \cdot 2 \equiv 6 \equiv -1 \qquad 3 \cdot 3 \equiv 9 \equiv 2$$

one has negative least absolute residue.

5.40.3

We will now give the above criterion another form. The least positive residues of $a,2a,\ldots,p'a$ modulo p are the integers

$$\beta_1,\beta_2,\ldots,\beta_{p'-\mu},\ p-\alpha_1,p-\alpha_2,\ldots,p-\alpha_\mu.$$

However we have*

$$1\cdot a = [\tfrac{a}{p}]p+r_1,\ 2\cdot a = [\tfrac{2a}{p}]+r_2,\ldots,p'\cdot a = [\tfrac{p'a}{p}]+r_{p'}$$

where $r_1,r_2,\ldots,r_{p'}$ are the least positive residues of $a,2a,\ldots,p'a$ modulo p. Adding these equalities we have

$$a(1+2+\ldots+p') = p([\tfrac{a}{p}]+\ldots+[\tfrac{p'a}{p}])+r_1+\ldots+r_{p'}$$

$$= pM + \sum_{j=1}^{p'-\mu} \beta_j + \sum_{i=1}^{\mu} (p-\alpha_i)$$

where $M = \sum_{k=1}^{p'} [\tfrac{ka}{p}]$. Consequently

$$a\cdot\frac{p'(p'+1)}{2} = pM+\Sigma\beta_j+ \Sigma\alpha_i+ \mu p - 2\Sigma\alpha_i$$

and thus, since $\Sigma\beta_j+ \Sigma\alpha_i= 1+2+\ldots+p'= p'(p'+1)/2$, we have

*For a real number x, [x] is the largest integer n satisfying $n \leqslant x$.

5.40.4

$$(a-1)\frac{p'(p'+1)}{2} \equiv pM + \mu p \quad (\mathrm{mod}\ 2)$$

$$\equiv -M + \mu \quad (\mathrm{mod}\ 2).$$

So

$$\mu \equiv (a-1)\frac{p'(p'+1)}{2} + M \quad (\mathrm{mod}\ 2).$$

Consequently

$$\left(\frac{a}{p}\right) = (-1)^{\mu} = (-1)^{(a-1)\left[p'(p'+1)/2\right]+M}. \tag{40.2}$$

As an application of this formula, we will calculate again $\left(\frac{-1}{p}\right)$ with its help. for $a = -1$

$$M = \left[\frac{-1}{p}\right]+\left[\frac{-2}{p}\right]+\ldots+\left[\frac{-p'}{p}\right] = -1-1-\ldots-1 = -p'.$$

So

$$\left(\frac{-1}{p}\right) = (-1)^{-2\left[p'(p'+1)/2\right]-p'} = \left[(-1)^{2}\right]^{-p'(p'+1)/2} \cdot (-1)^{-p'}$$

$$= 1 \cdot (-1)^{p'} = (-1)^{\frac{p-1}{2}}.$$

41. STUDY OF $\left(\frac{2}{p}\right)$ AND AN APPLICATION

According to formula (40.2) we have

$$\left(\frac{2}{p}\right) = (-1)^{\left[p'(p'+1)/2\right]+M}$$

where

$$M = \left[\frac{2}{p}\right]+\left[\frac{4}{p}\right]+\ldots+\left[\frac{2p'}{p}\right] = 0 + 0 + \ldots + 0 = 0$$

5.41.1

thus

$$\left(\frac{2}{p}\right) = (-1)^{p'(p'+1)/2} = (-1)^{\frac{p-1}{2} \cdot \frac{p+1}{2} \cdot \frac{1}{2}}$$

$$= (-1)^{\frac{p^2-1}{8}}. \tag{41.1}$$

We set

$$p = 8k+r \quad \text{where } r = \pm1,\pm3.$$

So

$$\frac{p^2-1}{8} = \frac{64k^2 + 16kr + r^2 - 1}{8} = 8k^2 + 2kr + \frac{r^2-1}{8}$$

$$\equiv \frac{r^2-1}{8} \pmod 2.$$

Thus

$$\left(\frac{2}{p}\right) = \begin{cases} 1 & \text{for } p \equiv r \equiv \pm1 \pmod 8 \\ -1 & \text{for } p \equiv r \equiv \pm3 \pmod 8. \end{cases} \tag{41.2}$$

<u>PROPOSITION 1</u> The odd prime divisors of $x^2 - 2$ (with integer x) have the form $8k\pm1$ with integer k.

    <u>Proof</u> If $p|x^2 - 2$ then 2 is a quadratic residue mod p and thus $\left(\frac{2}{p}\right) = 1$. Thus, according to (41.2) we have

$$p \equiv \pm1 \pmod 8 \quad \text{so} \quad p = 8k\pm1.$$

<u>PROPOSITION 2</u> The odd prime divisors of $x^2 + 2$ (with integer x) have the form $8k+1$ or $8k+3$.

    <u>Proof</u> We have $p|x^2 + 2 \Rightarrow \left(\frac{-2}{p}\right) = 1$. However

$$\left(\frac{-2}{p}\right) = \left(\frac{-1}{p}\right)\left(\frac{2}{p}\right)$$

5.41.2

and

$$\left(\frac{2}{p}\right) = \begin{cases} 1 & \text{for} \quad p \equiv \pm 1 \quad (\text{mod } 8) \\ -1 & \text{for} \quad p \equiv \pm 3 \quad (\text{mod } 8) \end{cases}$$

and

$$\left(\frac{-1}{p}\right) = \begin{cases} 1 & \text{for} \quad p \equiv 1 \quad (\text{mod } 4) \iff p \equiv 1,-3 \quad (\text{mod } 8) \\ -1 & \text{for} \quad p \equiv 3 \quad (\text{mod } 4) \iff p \equiv 3,-1 \quad (\text{mod } 8). \end{cases}$$

So

$$\left(\frac{-2}{p}\right) = \begin{cases} 1 & \text{for} \quad p \equiv 1,3 \qquad\qquad (\text{mod } 8) \\ -1 & \text{for} \quad p \equiv -1,-3, \equiv 7,5 \quad (\text{mod } 8). \end{cases}$$

which is what we wished to prove.

42. THE LAW OF QUADRATIC RECIPROCITY

From the formula (40.2) it follows for odd a that

$$\left(\frac{a}{p}\right) = (-1)^M \quad \text{where} \quad M = \sum_{n=1}^{p'} \left[\frac{na}{p}\right].$$

Let now $a = q$, an odd prime distinct from p. We will have

$$\left(\frac{q}{p}\right) = (-1)^M \quad \text{with} \quad M = \sum_{n=1}^{p'} \left[\frac{nq}{p}\right], \quad p' = \frac{p-1}{2}$$

$$\left(\frac{p}{q}\right) = (-1)^N \quad \text{with} \quad N = \sum_{n=1}^{q'} \left[\frac{np}{q}\right], \quad q' = \frac{q-1}{2}. \tag{42.1}$$

The Law of Quadratic Reciprocity is expressed by the relation

$$M+N = p'q' = \frac{p-1}{2}\,\frac{q-1}{2}. \tag{42.2}$$

5.42.1

PROOF OF THE RELATION

In the xy plane we consider the points (x,y) with integer coordinates $1 \leqslant x \leqslant p' = \frac{p-1}{2}$ and $1 \leqslant y \leqslant q' = \frac{q-1}{2}$. We will call such points lattice points. The number of these is $p'q'$. We also consider the straight line with equation

$$\frac{y}{x} = \frac{q}{p} .$$

None of the above $p'q'$ lattice points (x,y) lies on this straight line. Actually let a lattice point (x_1,y_1) satisfy the equation of the line;

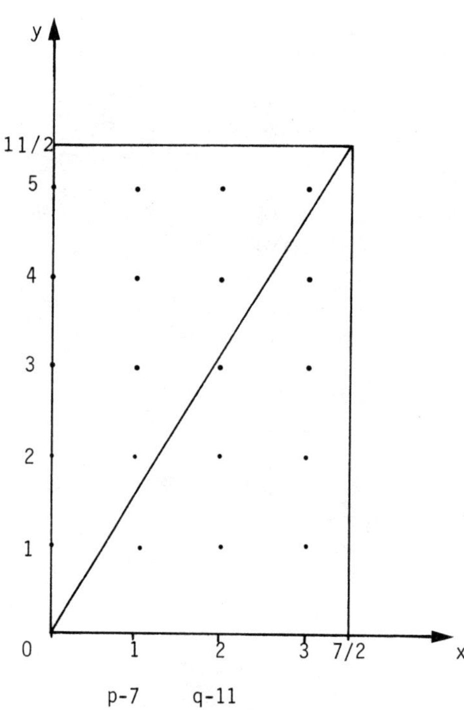

$y_1/x_1 = q/p$. Then $py_1 = qx_1$ so, since $(q,p) = 1$, $p|x_1$ and $q|y_1$. But this contradicts $1 \leqslant x_1 \leqslant p' < p$ and $1 \leqslant y_1 \leqslant q' < q$. Let us now count the number of lattice points (x,y) which are found below the line $y = (q/p)x$. For a given integral value n of x $(1 \leqslant n \leqslant p')$ the corresponding y will satisfy $1 \leqslant y < (q/p)n$ and, since the y are integers, $1 \leqslant y \leqslant [\frac{nq}{p}]$. So the total number of lattice points below the straight line $\frac{y}{x} = \frac{q}{p}$ is equal to $\sum_{n=1}^{p'} [\frac{nq}{p}] = M$.

5.42.2

It is proved in the same way that the number of lattice points above the line $\frac{y}{x} = \frac{q}{p}$ is $\sum_{n=1}^{q'} [\frac{np}{q}] = N$. But the total number, above and below, of lattice points is $p'q'$. Therefore

$$M+N = p'q' = \frac{p-1}{2} \frac{q-1}{2}.$$

From (42.1) and (42.2) we have immediately

<u>LAW OF QUADRATIC RECIPROCITY</u>

$$\left(\frac{p}{q}\right)\left(\frac{q}{p}\right) = (-1)^{\frac{p-1}{2} \cdot \frac{q-1}{2}} \tag{42.3}$$

for distinct odd primes p and q.

Since $\left(\frac{q}{p}\right) = \pm 1$, $\left(\frac{q}{p}\right)^2 = 1$, so multiplying both sides of (42.3) by $\left(\frac{q}{p}\right)$, we have

$$\left(\frac{p}{q}\right) = (-1)^{\frac{p-1}{2} \frac{q-1}{2}} \left(\frac{q}{p}\right). \tag{42.4}$$

Accordingly

$\left(\frac{p}{q}\right) = \left(\frac{q}{p}\right)$ if p or q is of the form 4k+1 with integer k

$\left(\frac{p}{q}\right) = -\left(\frac{q}{p}\right)$ if p and q are both of the form 4k+3 with integer k.

<u>EXAMPLES</u>. $\left(\frac{19}{23}\right) = -\left(\frac{23}{19}\right) = -\left(\frac{4}{19}\right) = -\left(\frac{2}{19}\right) \cdot \left(\frac{2}{19}\right) = -1.$

$$\left(\frac{13}{23}\right) = \left(\frac{23}{13}\right) = \left(\frac{10}{13}\right) = \left(\frac{2}{13}\right)\left(\frac{5}{13}\right) = (-1)^{(13^2 - 1)/8} \left(\frac{13}{5}\right)$$

$$= (-1)\left(\frac{3}{5}\right) = (-1)\left(\frac{5}{3}\right) = (-1)\left(\frac{2}{3}\right) = (-1)(-1)^{(3^2 - 1)/8} = (-1) \cdot (-1) = 1.$$

5.43.1

43. DETERMINATION OF THE ODD PRIMES p FOR WHICH $(\frac{a}{p}) = 1$ WITH GIVEN q

  <u>PROBLEM</u> To determine all the odd primes p for which $(\frac{q}{p}) = 1$ with a given odd prime q.

  <u>Solution</u> Let first q = 4n+1. Then

  $$(\frac{q}{p}) = (\frac{p}{q}).$$

  Our problem then reduces to the following: determine all the p for which $(\frac{p}{q}) = 1$. In other words, we must find all the p which are quadratic residues mod q. As we saw in §33, these residues are integers a_j which satisfy $a^{(q-1)/2} \equiv 1$ (mod q). We now determine the integers a_j from the set $\{1,2,\ldots,q-1\}$ which are solutions of the congruence $a^{(q-1)/2} \equiv 1$ (mod q). The required p are the odd primes which belong to the arithmetic progressions a_j+kq with k ϵ $\mathbb{Z}$.

  Now let q = 4n+3. Now we must find all the p of the form 4k+1 for which $(\frac{p}{q}) = 1$ and all the p of the form 4k+3 for which $(\frac{p}{q}) = -1$. Let a_j be the integers from the set $\{1,2,\ldots,q-1\}$ satisfying $a_j^{(q-1)/2} \equiv 1$ (mod q) and b_i the integers from the same set satisfying $b_i^{(q-1)/2} \equiv -1$ (mod q). The required p are the primes of the form 4k+1 belonging to the arithmetic progressions $a_j+\ell q$, and the primes of the form 4k+3 belonging to the arithmetic progressions $b_i+\ell q$, where ℓ ϵ $\mathbb{Z}$.

<u>EXAMPLE</u> Let q = 13. Since 13 = 4n+1, we have

$$(\frac{13}{p}) = (\frac{p}{13}).$$

We determine the a for which $a^{(p-1)/2} \equiv a^6 \equiv 1$ (mod 13). So we have to solve the linear congruence with respect to Ind a

5.43.2

$$6 \text{ Ind } a \equiv \text{Ind } 1 = 12 \quad (\text{mod } 12)$$
$$\text{Ind } a \equiv 2 \quad (\text{mod } 2)$$
$$\equiv 2,4,6,8,10,12 \quad (\text{mod } 12).$$

The corresponding a are (see table, §28)

$$a \equiv 4,3,12,9,10,1 \ (\text{mod } 13).$$

The required p are then the odd primes which belong to the following six

arithmetic progressions.

$$4+13k, \ 3+13k, \ 12+13k, \ 9+13k, \ 10+13k, \ 1+13k, \text{ with } k \in \mathbb{Z}.$$

For example 17 and 43 belong to the 1st progression; 29 and 107 to the

second; 103, 181 to the third; 61 to the fourth; 23 to the fifth; and 53

to the 6th. We verify for 23:

$$\left(\tfrac{13}{23}\right) = \left(\tfrac{23}{13}\right) = \left(\tfrac{-3}{13}\right) \equiv (-3)^6 \equiv 9^3 \equiv (-4)^3 = -64 \equiv 1 \quad (\text{mod } 13).$$

44. GENERALIZATION OF THE LEGENDRE SYMBOL $\left(\tfrac{a}{p}\right)$ TO THE JACOBI SYMBOL

We will represent with P an arbitrary odd natural number and with a an

arbitrary integer relatively prime to P: $(a,P) = 1$. We now give the

following definition of the Jacobi symbol:

<u>DEFINITION</u> $\left(\tfrac{a}{P}\right) = 1$ when $P = 1$ and a any integer. For $P > 1$

$$\left(\tfrac{a}{P}\right) = \left(\tfrac{a}{p_1}\right)\cdots\left(\tfrac{a}{p_k}\right) \text{ for } P = p_1 p_2 \cdots p_k \text{ with } k \geqslant 1.$$

where p_j, $(j = 1,\ldots,k)$ are odd primes (not necessarily distinct) and

$(a,P) = 1$.

As we see

$$\left(\tfrac{a}{P}\right) = \left(\tfrac{a}{p}\right) \text{ when } P = p.$$

<u>Properties of the Jacobi symbol</u> $\left(\tfrac{a}{P}\right)$:

<u>THEOREM 1</u> $\left(\tfrac{a}{P}\right) = \pm 1$.

The proof is immediate.

5.44.1

<u>THEOREM 2</u> $(\frac{a}{P}) = 1$ when $a = m^2$ is the square of an integer (and P is arbitrary) or when $P = n^2$ is the square of an odd integer (and a is arbitrary).

<u>Proof</u> $(\frac{m^2}{1}) = 1$ and for $P = p_1 \ldots p_k$ we have

$$(\frac{m^2}{P}) = (\frac{m^2}{p_1}) \ldots (\frac{m^2}{p_k}) = 1 \ldots 1 = 1.$$

In the second case, $P = (p_1 \ldots p_k)^2$ then

$$(\frac{a}{P}) = (\frac{a}{p_1})^2 \ldots (\frac{a}{p_k})^2 = 1 \ldots 1 = 1.$$

<u>THEOREM 3</u> $(\frac{a}{P}) = (\frac{b}{P})$ when $a \equiv b \pmod{P}$.

<u>Proof</u> This is clear when $P = 1$. For $P = p_1 \ldots p_k$ the hypothesis $a \equiv b \pmod{P}$ implies $a \equiv b \pmod{p_j}$, for $j = 1, \ldots, k$, and this implies

$$(\frac{a}{p_j}) = (\frac{b}{p_j}) \text{ for } j = 1, \ldots, k$$

so

$$(\frac{a}{P}) = (\frac{b}{P}).$$

<u>THEOREM 4</u> $(\frac{ab}{P}) = (\frac{a}{P})(\frac{b}{P})$.

<u>Proof</u> This is clear when $P = 1$. For $P = p_1 \ldots p_k$, recall that

$$(\frac{ab}{p_j}) = (\frac{a}{p_j})(\frac{b}{p_j}) \qquad j = 1, \ldots, k.$$

5.44.2

so

$$\left(\frac{ab}{P}\right) = \prod_{j=1}^{k} \left(\frac{ab}{p_j}\right) = \prod_{j=1}^{k} \left(\frac{a}{p_j}\right) \prod_{j=1}^{k} \left(\frac{b}{p_j}\right) = \left(\frac{a}{P}\right)\left(\frac{b}{P}\right).$$

<u>THEOREM 5</u> $\left(\frac{a}{PQ}\right) = \left(\frac{a}{P}\right)\left(\frac{a}{Q}\right).$

<u>Proof</u> This is clear when $P = 1$ or $Q = 1$. For $P = p_1 \ldots p_k$ and $Q = q_1 \ldots q_\ell$ we have $PQ = p_1 \ldots p_k q_1 \ldots q_\ell$, so

$$\left(\frac{a}{PQ}\right) = \left(\frac{a}{p_1}\right)\ldots\left(\frac{a}{p_k}\right)\left(\frac{a}{q_1}\right)\ldots\left(\frac{a}{q_\ell}\right) = \left(\frac{a}{P}\right)\left(\frac{a}{Q}\right).$$

<u>THEOREM 6</u> $\left(\frac{-1}{P}\right) = (-1)^{(P-1)/2}.$

<u>Proof</u> This is clear for $P = 1$. For $P = p_1 \ldots p_k$ we have

$$\left(\frac{-1}{P}\right) = \prod_{j=1}^{k} \left(\frac{-1}{p_j}\right) = \prod_{j=1}^{k} (-1)^{(p_j - 1)/2} = (-1)^{\sum_{j=1}^{k} (p_j - 1)/2}.$$

However, we have $\sum_{j=1}^{k} \frac{p_j - 1}{2} \equiv \frac{P-1}{2}$ (mod 2), since, when $k=1$ we have $\frac{p_1 - 1}{2} = \frac{P-1}{2}$ and when $k \geqslant 2$ we have

$$P = (1+(p_1 - 1))(1+(p_2 - 1))\ldots(1+(p_k - 1))$$

$$= 1 + \sum_{j=1}^{k}(p_j - 1) + 4A \text{ with integer } A,$$

because, since $p_j - 1$ is even, any product of 2 or more factors of the form $p_j - 1$ will be divisible by 4.

So

$$\frac{P-1}{2} \equiv \frac{1}{2} \sum_{j=1}^{k} (p_j - 1) \quad (\text{mod } 2)$$

5.44.3

and thus

$$(-1)^{(P-1)/2} = (-1)^{\sum\limits_{j=1}^{k} (p_j - 1)/2} = \left(\frac{-1}{P}\right)$$

COROLLARY 1

$$\left(\frac{-1}{P}\right) = \begin{cases} 1 & \text{for } P = 4n+1 \\ -1 & \text{for } P = 4n+3 \end{cases} \qquad \text{with integer } n \geqslant 0.$$

THEOREM 7 $\left(\frac{2}{P}\right) = (-1)^{(P^2 - 1)/8}$.

Proof This is clear for $P = 1$ or for $P = p$. For $P = p_1 \ldots p_k$, $k \geqslant 2$ we have

$$\left(\frac{2}{P}\right) = \left(\frac{2}{p_1}\right) \ldots \left(\frac{2}{p_k}\right) = (-1)^{(p_1^2 - 1)/8} \ldots (-1)^{(p_k^2 - 1)/8}$$

$$= (-1)^{\sum\limits_{j=1}^{k} (p_j^2 - 1)/8}.$$

However, we have

$$P^2 = (1+(p_1^2 - 1))\ldots(1+(p_k^2 - 1))$$

$$= 1 + \sum\limits_{j=1}^{k} (p_j^2 - 1) + 64A \qquad \text{with integer } A,$$

because, since p_j is odd, $p_j^2 - 1 \equiv 0 \pmod 8$ and a product of two or more factors of the form $p_j^2 - 1$ will be divisible by 64.

5.44.4

Thus

$$\frac{P^2-1}{8} = \sum_{j=1}^{k} \frac{p_j^2-1}{8} + 8A$$

$$\equiv \sum_{j=1}^{k} \frac{p_j^2-1}{8} \quad (\mathrm{mod}\ 2)$$

so

$$\left(\frac{2}{P}\right) = (-1)^{\frac{P^2-1}{8}}.$$

COROLLARY 2 $\quad \left(\frac{2}{P}\right) = \begin{cases} 1 & \text{for } P = 8h\pm1 \\ -1 & \text{for } P = 8h\pm3 \end{cases}$ with integer h.

Proof Actually, if $P = 8h\pm1$ we have

$$P^2 = (8h\pm1)^2 = 64h^2 \pm 16h+1$$

so

$$\frac{P^2-1}{8} = 8h^2 \pm 2h = 2(4h^2 \pm h) = 2A \text{ with integer } A.$$

Thus

$$\left(\frac{2}{P}\right) = (-1)^{2A} = 1.$$

If $P = 8h\pm3$ we have

$$P^2 = (8h\pm3)^2 = 64h^2 \pm 48h+9$$

so

$$\frac{P^2-1}{8} = 8h^2 \pm 6h + 1 = 2(4h^2 \pm 3h) + 1 = 2A + 1$$

and thus

$$\left(\frac{2}{P}\right) = (-1)^{2A+1} = -1.$$

5.44.5

COROLLARY 3 $\left(\frac{-2}{P}\right) = \begin{cases} 1 & \text{for } P \equiv 1,3 \pmod 8 \\ -1 & \text{for } P \equiv -1,-3 \pmod 8. \end{cases}$

Proof From Theorem 4 we have $\left(\frac{-2}{P}\right) = \left(\frac{-1}{P}\right)\left(\frac{2}{P}\right)$ and it suffices to make use of Corollaries 1 and 2*.

THEOREM 8 $\left(\frac{P}{Q}\right)\left(\frac{Q}{P}\right) = (-1)^{\frac{P-1}{2}\frac{Q-1}{2}}$ where P and Q are odd natural numbers and $(P,Q) = 1$.

Proof This is clear when $P = 1$ or $Q = 1$. For $P = p_1 \ldots p_k$ and $Q = q_1 \ldots q_\ell$ we have, according to the definition and theorem 4,

$$\left(\frac{P}{Q}\right) = \left(\frac{P}{q_1}\right)\ldots\left(\frac{P}{q_\ell}\right) = \prod_{\substack{i=1,\ldots,k \\ j=1,\ldots,\ell}} \left(\frac{p_i}{q_j}\right)$$

and

$$\left(\frac{Q}{P}\right) = \prod_{\substack{i=1,\ldots,k \\ j=1,\ldots,\ell}} \left(\frac{q_j}{p_i}\right).$$

So

$$\left(\frac{P}{Q}\right)\left(\frac{Q}{P}\right) = \prod_{\substack{i=1,\ldots,k \\ j=1,\ldots,\ell}} \left(\frac{p_i}{q_j}\right)\left(\frac{q_j}{p_i}\right) = \prod_{\substack{i=1,\ldots,k \\ j=1,\ldots,\ell}} (-1)^{\frac{p_i-1}{2}\frac{q_j-1}{2}}$$

$$= (-1)^{\sum_{i,j} \frac{p_i-1}{2}\frac{q_j-1}{2}}.$$

*See proof of Proposition 2 of §41.

5.44.6

$$= (-1)^{\left(\sum\limits_i \frac{p_i - 1}{2}\right)\left(\sum\limits_j \frac{q_j - 1}{2}\right)}.$$

However, as we have seen in the proof of Theorem 6

$$\frac{P-1}{2} \equiv \sum\limits_{i=1}^{k} \frac{p_i - 1}{2} \quad \text{and} \quad \frac{Q-1}{2} \equiv \sum\limits_{j=1}^{\ell} \frac{q_j - 1}{2} \qquad (\text{mod } 2),$$

so $\left(\sum\limits_{i=1}^{k} \frac{p_i - 1}{2}\right)\left(\sum\limits_{j=1}^{\ell} \frac{q_j - 1}{2}\right) \equiv \frac{P-1}{2} \cdot \frac{Q-1}{2} \quad (\text{mod } 2)$

and $\left(\frac{P}{Q}\right)\left(\frac{Q}{P}\right) = (-1)^{\frac{P-1}{2} \cdot \frac{Q-1}{2}}.$

$\underline{\text{COROLLARY 4}} \quad \left(\frac{P}{Q}\right)\left(\frac{Q}{P}\right) = \begin{cases} 1 & \text{if } P \equiv 1 \text{ or } Q \equiv 1 \quad (\text{mod } 4) \\ -1 & \text{if } P \equiv 3 \text{ and } Q \equiv 3 \quad (\text{mod } 4) \end{cases}$

and thus $\left(\frac{P}{Q}\right) = \left(\frac{Q}{P}\right)$ when $P \equiv 1$ or $Q \equiv 1$ (mod 4)

and $\left(\frac{P}{Q}\right) = -\left(\frac{Q}{P}\right)$ when $P \equiv 3$ and $Q \equiv 3$ (mod 4).

The proof is immediate.

From the above we conclude

$$\left(\frac{-P}{Q}\right) = \left(\frac{-1}{Q}\right)\left(\frac{P}{Q}\right) = (-1)^{(Q-1)/2}(-1)^{(P-1)/2 \cdot (Q-1)/2} \left(\frac{Q}{P}\right)$$
$$= (-1)^{(P+1)/2 \cdot (Q-1)/2} \left(\frac{Q}{P}\right).$$

Hence, when $P \equiv 3 \equiv -1$ (mod 4) then $(P+1)/2 \equiv 0$ (mod 2) and thus

$$\left(\frac{-P}{Q}\right) = \left(\frac{Q}{P}\right) \text{ when } P \equiv 3 \equiv -1 \text{ (mod 4)}.$$

Since $\left(\frac{P}{Q}\right) = \left(\frac{Q}{P}\right)$ when $P \equiv 1$ (mod 4) we have the formula

$$P \equiv \pm 1 \text{ (mod 4)} \Rightarrow \left(\frac{\pm P}{Q}\right) = \left(\frac{Q}{P}\right) \tag{44.1}$$

where the + sign corresponds to the + sign and the - to the -.

$\underline{\text{EXAMPLE}}$ Calculation of $\left(\frac{566}{1409}\right)$.

$$\left(\frac{566}{1409}\right) = \left(\frac{2}{1409}\right)\left(\frac{283}{1409}\right) = 1\left(\frac{283}{1409}\right) \quad \text{since } 1409 \equiv 1 \text{ (mod 8)}$$

5.44.7

$$= \left(\tfrac{1409}{283}\right) = \left(\tfrac{-6}{283}\right) \qquad \text{since } 1415 = 5 \cdot 283$$

$$= \left(\tfrac{-1}{283}\right)\left(\tfrac{2}{283}\right)\left(\tfrac{3}{283}\right)$$

$$= (-1)(-1)\left(\tfrac{3}{283}\right) \qquad \text{since } 283 \equiv 3 \ (\mathrm{mod}\ 4) \text{ and } 283 \equiv 3 \ (\mathrm{mod}\ 8)$$

$$= \left(\tfrac{3}{283}\right) = (-1)\left(\tfrac{283}{3}\right) \qquad \text{since } 283 \equiv 3 \ (\mathrm{mod}\ 4)$$

$$= (-1)\left(\tfrac{1}{3}\right) = (-1)\cdot 1 = -1.$$

45. COMPLETION OF THE SOLUTION OF THE SECOND PROBLEM OF §32

We wish to determine all the odd primes p for which $(a,p) = 1$ and $\left(\tfrac{a}{p}\right) = 1$ with given a.

We generalize the problem to finding all the odd natural numbers P for which $(a,P) = 1$ and $\left(\tfrac{a}{P}\right) = 1$ with given a.

AUXILIARY PROBLEM Let $A > 0$ be an odd integer. We wish to determine all the natural numbers P for which a) $(A,P) = 1$ and $\left(\tfrac{P}{A}\right) = 1$ and b) $(A,P) = 1$ and $\left(\tfrac{P}{A}\right) = -1$. This may be done as follows:

Let $\{k_1,\ldots,k_{\phi(A)}\}$ be a reduced system of residues mod A. We calculate $\left(\tfrac{k_j}{A}\right)$, $j = 1,\ldots,\phi(A)$. The k_j for which $\left(\tfrac{k_j}{A}\right) = 1$ we designate with r, the rest with s; then we have

$$\left(\tfrac{P}{A}\right) = \begin{cases} 1 & \text{when } P \equiv r \ (\mathrm{mod}\ A) \\ -1 & \text{when } P \equiv s \ (\mathrm{mod}\ A). \end{cases}$$

We ask: how many r are there, and how many s. We distinguish two cases.

I) A is the square of an integer.

Then, according to theorem 2 of §44, $\left(\tfrac{k_j}{A}\right) = 1$ for $j = 1,\ldots,\phi(A)$. So the number of r is $\phi(A)$, and the number of s is 0.

5.45.1

II) A is not the square of an integer. Then

$$\sum_{j=1}^{\phi(A)} \left(\frac{k_j}{A}\right) = A_r - A_s = S$$

where A_r is the number of r, and A_s the number of s. Now let k be relatively prime to A. Then the set $\{kk_1, kk_2, \ldots, kk_{\phi(A)}\}$ coincides with the set $\{k_1, \ldots, k_{\phi(A)}\}$ and thus

$$\left(\frac{k}{A}\right)S = \left(\frac{k}{A}\right)\sum_{j=1}^{\phi(A)} \left(\frac{k_j}{A}\right) = \sum_{j=1}^{\phi(A)} \left(\frac{kk_j}{A}\right) = \sum_{j=1}^{\phi(A)} \left(\frac{k_j}{A}\right) = S$$

so

$$\left(\frac{k}{A}\right)S = S.$$

We will now show how this relation implies that $S = 0$. Actually, since A is not the square of an integer, we must have $A = p^{2\alpha+1}A'$ where $\alpha \geqslant 0$ and A' is an integer with $(A',p) = 1$, and p is an odd prime. We will use this p to show that there exists a k satisfying $\left(\frac{k}{A}\right) = -1$.

Now there certainly exists a β satisfying $\left(\frac{\beta}{p}\right) = -1$, for, according to §33 there exist $(p-1)/2$ quadratic non-residues mod p. Selecting one such, we next determine a k_0 which satisfies the system

$$k_0 \equiv \beta \pmod{p} \qquad\qquad k_0 \equiv 1 \pmod{A'}.$$

The system has solutions, since $(p,A') = 1$. For the solution k_0, we have $(k_0,p) = (\beta,p) = 1$ and $(k_0,A') = (1,A') = 1$. We have

$$\left(\frac{k_0}{A}\right) = \left(\frac{k_0}{p^{2\alpha+1}}\right)\left(\frac{k_0}{A'}\right) = \left(\frac{k_0}{p}\right)^{2\alpha+1}\left(\frac{k_0}{A'}\right) = \left(\frac{\beta}{p}\right)^{2\alpha+1}\left(\frac{1}{A'}\right) = (-1)^{2\alpha+1}\cdot 1 = -1.$$

5.45.2

We introduce this value k_0 into the equation $(\frac{k}{A})S = S$ and find

$$-S = (\frac{k_0}{A})S = S$$

from which we conclude that $S = 0$ and thus $A_r = A_s$. Thus in case II (A not the square of an integer) the r and s are equal in number.

We return now to the original problem: Given an integer a, we seek those odd integers $P > 0$ for which $(a,P) = 1$ and $(\frac{a}{P}) = 1$.

Without loss of generality we may assume that a has no divisors which are squares of integers. Indeed, if $a = n^2 a'$, then

$$(\frac{a}{P}) = (\frac{n^2}{P})(\frac{a'}{P}) = 1 \cdot (\frac{a'}{P})$$

where $a' = \pm 1$ or $a' = \pm p_1,\ldots,p_k$ with primes p_i satisfying $p_i \neq p_j$ if $i \neq j$.

If $a' = 1$ then $(\frac{a}{P}) = 1$ for any odd natural number P.

If $a' = -1$ then

$(\frac{a}{P}) = 1$ for any odd natural number $P \equiv 1$ (mod 4)

and

$(\frac{a}{P}) = -1$ for any odd natural number $P \equiv -1$ (mod 4).

If $a' = \pm p_1 \ldots p_k$ with $p_i \neq p_j$ for $i \neq j$, we determine the P for which $(a',P) = 1$ and $(\frac{a'}{P}) = 1$, and afterward remove those P which are not relatively prime to a, that is, the P with $(a,P) > 1$.

Let now a be of the form $a = p_1 \ldots p_k$ with $p_i \neq p_j$ for $i \neq j$. We distinguish two cases: I) a is odd and II) a is even. In each case there are two possible subcases. We set $a = \pm A$, with A a positive integer.

5.45.3

<u>CASE I</u> a is odd.

1st SUBCASE $a \equiv 1 \pmod 4$. From $a = \pm A \equiv 1 \pmod 4$ where A is a natural number it follows that $A \equiv \pm 1 \pmod 4$. According to (44.1) we will have

$$\left(\tfrac{\pm A}{P}\right) = \left(\tfrac{P}{A}\right) \text{ so } \left(\tfrac{a}{P}\right) = \left(\tfrac{P}{A}\right).$$

Thus, by the auxiliary problem

$$\left(\tfrac{a}{P}\right) = 1 \text{ when } P \equiv r \pmod A \text{ with P odd.}$$

Since A is not the square of an integer, there are $\tfrac{1}{2}\phi(A)$ residues r. The required P thus belong to $\tfrac{1}{2}\phi(A)$ arithmetic progressions, each with difference 2A. Actually, since

$$\left(\tfrac{P}{A}\right) = \left(\tfrac{a}{P}\right) = 1$$

we will have $P \equiv r \pmod A$ so

$$P = r+mA \text{ with } r+mA \text{ odd.}$$

Thus $r+mA \equiv 1 \pmod 2$ and, regarding this as a linear congruence with respect to m, there is a solution m_0 since the coefficient A of m is odd and so $(A,2) = 1$. With this m_0, we have

$m \equiv m_0 \pmod 2$ and $m = m_0 + 2\lambda$, $\lambda \in \mathbb{Z}$. Hence

$$P = r+mA = r+(m_0 + 2\lambda)A = r+m_0 A+\lambda\cdot 2A, \text{ with } \lambda \in \mathbb{Z}, \text{ and P belongs to}$$

an arithmetic progression with difference 2A.

2nd SUBCASE $a \equiv -1 \pmod 4$, then $-a \equiv 1 \pmod 4$ and, setting $-a = \pm A$ with A a natural number we have $\pm A \equiv 1 \pmod 4$. Then

$$\left(\tfrac{a}{P}\right) = \left(\tfrac{-1}{P}\right)\left(\tfrac{-a}{P}\right) = \left(\tfrac{-1}{P}\right)\left(\tfrac{\pm A}{P}\right) = \left(\tfrac{-1}{P}\right)\left(\tfrac{P}{A}\right)$$

so

$$\left(\tfrac{a}{P}\right) = 1 \text{ for } \begin{cases} P \equiv 1 \pmod 4 \text{ and } P \equiv r \pmod A \\ P \equiv -1 \pmod 4 \text{ and } P \equiv s \pmod A. \end{cases}$$

Hence the P are found in $\phi(A)$ distinct arithmetic progressions, each with difference 4A.

5.45.4

CASE II a even, and not divisible by the square of an integer. Again we have two cases.

1st SUBCASE $a = 2B'$, where B' is an odd integer and $B' \equiv 1$ (mod 4).

Setting $B' = \pm A'$ where A' is an odd positive integer, we have

$$\left(\frac{a}{P}\right) = \left(\frac{\pm 2A'}{P}\right) = \left(\frac{2}{P}\right)\left(\frac{\pm A'}{P}\right) = \left(\frac{2}{P}\right)\left(\frac{P}{A'}\right).$$

So

$$\left(\frac{a}{P}\right) = 1 \text{ when } \left(\frac{2}{P}\right) \text{ and } \left(\frac{P}{A'}\right) \text{ are both } +1 \text{ or both } -1.$$

We have $\left(\frac{2}{P}\right)$ and $\left(\frac{P}{A'}\right)$ simultaneously equal to 1 when $P \equiv \pm 1$ (mod 8) and $P \equiv r$ (mod A'). We have $\left(\frac{2}{P}\right)$ and $\left(\frac{P}{A'}\right)$ simultaneously equal to -1 when $P \equiv \pm 3$ (mod 8) and $P \equiv s$ (mod A'). Hence the required P belong to $2\phi(A')$ arithmetic progressions with differences $8A'$.

2nd SUBCASE $a = 2B'$ where B' is an odd integer and $B' \equiv -1$ (mod 4).

Then $-B' \equiv 1$ (mod 4). We set $-B' = \pm A'$ with A' an odd positive integer. Then

$$\left(\frac{a}{P}\right) = \left(\frac{2B'}{P}\right) = \left(\frac{-2}{P}\right)\left(\frac{-B'}{P}\right) = \left(\frac{-2}{P}\right)\left(\frac{\pm A'}{P}\right) = \left(\frac{-2}{P}\right)\left(\frac{P}{A'}\right)$$

$$= 1 \begin{cases} \text{for } P \equiv 1,3 \text{ (mod 8) and } P \equiv r \text{ (mod } A') \\ \text{for } P \equiv -1,-3 \text{ (mod 8) and } P \equiv s \text{ (mod } A'). \end{cases}$$

Thus the required P belong to $2\phi(A')$ arithmetic progressions with differences $8A'$.

5.P.1

PROBLEMS FOR CHAPTER 5

1. Verify that the criterion of Euler is a special case of Theorem 1 of §30.

2. Calculate $\left(\frac{2}{29}\right)$ and $\left(\frac{3}{31}\right)$ using Euler's criterion.

3. Using the technique of §34, prove the following generalization of the
    proposition of that section.

    Let $(n,p) = 1$ and $(a,p) = 1$. For $x^n \equiv a \pmod{p^r}$ to be solvable it
    is necessary and sufficient that $x^n \equiv a \pmod{p}$ be solvable.

    While proving this, note that a solution mod p^e generates a solution
    mod p^{2e}. This enormously improves the efficiency of the algorithm.

4. Taking note of the final remark in Problem 3, solve whichever of the
    following equations are solvable
    a) $x^2 \equiv 114 \pmod{5^4}$,   b) $x^2 \equiv 767 \pmod{5^4}$,
    c) $x^2 \equiv 10685 \pmod{29^3}$,   d) $x^2 \equiv 3229 \pmod{3^8}$.

5. Using the technique of problem 3, solve
    a) $x^4 \equiv 171 \pmod{5^4}$,   b) $x^3 \equiv 22743 \pmod{29^3}$.

6. Show that x_0, x_0+2^{k-1}, $-x_0$, $-x_0+2^{k-1}$, where x_0 is odd, are all
    incongruent mod 2^k, for $k \geqslant 3$.

7. The proof of the proposition in §35 may be restructured into the
    following algorithm: Assume $a \equiv 1 \pmod 8$. Given a solution x_1
    of $x^2 \equiv a \pmod{2^k}$, form $\dfrac{a-x_1^2}{2^k} = n$.
    If n is even, $n = 2^s n'$, n' odd then x_1 is a solution of
    $x^2 \equiv a \pmod{2^{k+s}}$.

    If n is odd then x_1+2^{k-1} is a solution of $x^2 \equiv a \pmod{2^{k+1}}$.

    The process is repeated until the desired exponent of 2 is attained.

5.P.2

a) Apply the algorithm of Problem 7 to the solution of

   A) $x^2 \equiv 769 \pmod{2^{10}}$ B) $x^2 \equiv 761 \pmod{2^{10}}$.

b) Construct a proof of the existence of solutions of $x^2 \equiv a \pmod{2^e}$

   where $e \geqslant 3$ and $a \equiv 1 \pmod 8$ based on the algorithm of Problem 7.

8. Show that, for $k \geqslant 2$, $\omega = 1 + 2^{k-1}$ is a primitive cube root of unity

   mod 3^k. (This means 1, ω, ω^2, are all the cube roots of unity mod 3^k.)

9. a) Construct an algorithm similar to that in Problem 7 for solving

   $$x^3 \equiv a \pmod{3^k} \text{ where } (a,3) = 1$$

   and use it to prove

   $x^3 \equiv a \pmod{3^k}$ is solvable for $k > 2$ if and only if $a \equiv 1, -1 \pmod 9$.

   Hint: The congruence class of $\dfrac{a - x^3}{3^k}$ is useful in determining the next

   step in the algorithm.

   b) Apply your algorithm to the problem $x^3 \equiv 422 \pmod{3^7}$. Then find all

   solutions using problem 8.

10. A. Cayley observed that it is unnecessary to know the value of $\left(\frac{2}{P}\right)$, since

    if a is even, $\left(\frac{a}{P}\right) = \left(\frac{a-P}{P}\right)$, where $a-P$ is odd. Use this technique to

    calculate $\left(\frac{365}{1847}\right)$, and compare it with the ordinary method. (This example

    is due to R. Dedekind, 1880.)

11. Find all solutions to those of the following congruences which are

    solvable, using the Chinese remainder theorem.

    a) $x^2 \equiv 14 \pmod{35}$,   b) $x^2 \equiv 35 \pmod{77}$,   c) $x^2 \equiv 254 \pmod{385}$.

    d) $x^2 \equiv 63 \pmod{140}$ e) $x^2 \equiv 89 \pmod{440}$.

In problems 12 through 15, p is an odd prime and we set $\left(\frac{a}{p}\right) = 0$ if $p \mid a$. The

symbol $\sum\limits_{x \bmod p}$ will indicate a sum where x runs through a complete system of

residues mod p, for example $\{0, 1, \ldots, p-1\}$. In this same context, the symbol

5.P.3

$\sum\limits_{x \not\equiv 0}$ will mean the sum over a reduced system of residues mod p.

12. Show that $\sum\limits_{x \bmod p} \left(\frac{x}{p}\right) = 0.$

13. Show that if $a \not\equiv 0 \pmod p$ then $\sum\limits_{x \bmod p} \left(\frac{ax+b}{p}\right) = 0.$
    What happens if $a \equiv 0 \pmod p$?

14. Show that $\sum\limits_{x \bmod p} \left(\frac{x(x+a)}{p}\right) = \begin{cases} -1 & \text{if } a \not\equiv 0 \pmod p \\ p-1 & \text{if } a \equiv 0 \pmod p \end{cases}.$
    Hint: For each $x \not\equiv 0 \pmod p$ let x' satisfy $xx' \equiv 1 \pmod p$.

    Then $\sum\limits_{x \not\equiv 0} \left(\frac{x(x+a)}{p}\right) = \sum\limits_{x \not\equiv 0} \left(\frac{xx'(xx'+ax')}{p}\right).$

15. Show that if $ax^2+bx+c \equiv 0 \pmod p$ is solvable and $a \not\equiv 0 \pmod p$ then

$$\sum\limits_{x \bmod p} \left(\frac{ax^2+bx+c}{p}\right) = \begin{cases} \left(\frac{a}{p}\right)(-1) & \text{if } p \nmid \Delta = b^2-4ac \\ \left(\frac{a}{p}\right)(p-1) & \text{if } p \mid \Delta = b^2-4ac. \end{cases}$$

(This result is true without the solvability assumption, but is then more

difficult to prove.) Hint: Factor and use 14.

In problems 16 through 19, $p \equiv 1 \pmod 4$ and the symbol $\sum\limits_{\pm x \bmod p}$ will mean
that x runs through a system containing exactly one element from each pair

$\{x,-x\}$ in a reduced system of residues mod p, for example $\{1,2,\ldots,(p-1)/2\}$.

16. Show $\sum\limits_{\pm x \bmod p} \left(\frac{x}{p}\right) = 0$.

17. Define $S(k)$ by $S(k) = \sum\limits_{x \bmod p} \left(\frac{x(x^2+k)}{p}\right).$

    a) Show $S(k) = 2\sum\limits_{\pm x \bmod p} \left(\frac{x(x^2+k)}{p}\right).$

    b) Show $S(kt^2) = \left(\frac{t}{p}\right)S(k)$.

    c) Let $a = \frac{1}{2}S(-1)$. Show $S(r)$ for any quadratic residue r (mod p) may be

    expressed in terms of a.

5.P.4

d) Show that $a \equiv -(\frac{2}{p})$ (mod 4).

Hint: Express a as $\sum\limits_{\pm x \bmod p} (\frac{x}{p})[(\frac{x^2-1}{p})-1]$ using problem 16.

Then show $a \equiv \sum\limits_{\pm x \bmod p} [(\frac{x^2-1}{p})-1]$ (mod 4). Use problem 15 to evaluate

the sum. Then compare to $(\frac{2}{p})$ for $p \equiv 1,5$ (mod 8).

18. Show that $\sum\limits_{k \not\equiv 0} (\frac{xy(x^2+k)(y^2+k)}{p}) = \begin{cases} -2\,(\frac{xy}{p}) & \text{for } y \not\equiv \pm x \text{ (mod } p) \\ (p-2)(\frac{xy}{p}) & \text{for } y \equiv \pm x \text{ (mod } p)\,. \end{cases}$

Hint: Reexpress as a $\sum\limits_{k \bmod p}$ and use problem 15. Note that k is the

variable.

19. Let $p' = (p-1)/2$, $(\frac{r}{p}) = +1$, $(\frac{n}{p}) = -1$.

a) Show $p'[S(r)]^2 = \sum\limits_{\pm t \bmod p} [S(rt^2)]^2$ and $p'[S(n)]^2 = \sum\limits_{\pm t \bmod p} [S(nt^2)]^2$.

b) Show $p'[S(r)]^2 + p'[S(n)]^2 = \sum\limits_{k \not\equiv 0} [S(k)]^2$.

c) Show that $p'[S(r)]^2 + p'[S(n)]^2 = 4pp'$.

Hint: Use b) and write the right side of b) as a triple sum

$\sum\limits_{k \not\equiv 0} \sum\limits_{x \not\equiv 0} \sum\limits_{y \not\equiv 0}$, rearrange with the k-sum innermost, and use

problem 18.

d) Conclude that p is a sum of two squares, and that $\frac{1}{2}S(n)$ is even (use

17d). [These results are due to Jacobsthal and Gorshkov.]

The following problem introduces a widely used notation useful for theoretical

as well as practical purposes.

20. For odd m, define $m^* = (-1)^{\frac{m-1}{2}} m$. Prove

a) $m^* = \begin{cases} m & \text{if } m \equiv 1 \pmod 4 \\ -m & \text{if } m \equiv -1 \pmod 4 \end{cases}$.

b) $m^* \equiv 1 \ (4)$.

c) $\frac{m^*+1}{2} \equiv 1 \pmod 2$.

d) $\frac{m^*-1}{4} \equiv \frac{m^2-1}{8} \pmod 2$.

Show

e) $(\frac{2}{p}) = (-1)^{\frac{p^*-1}{4}}$ for an odd prime p.

f) $(\frac{q}{p}) = (\frac{p^*}{q})$ for odd primes p,q.

Remark e) is somewhat more convenient for calculation than the form with

p^2.

We will now generalize the Jacobi symbol so that negative denominators are

permitted (problems 21, 22).

21. Define, for odd m,

a) $\chi_4(m) = (-1)^{\frac{m-1}{2}} \quad \chi_8(m) = (-1)^{\frac{m^*-1}{2}} \quad \chi_\infty(m) = (-1)^{\frac{\text{sgn } m-1}{2}}$.

Prove that, for odd m and n, $\chi_i(mn) = \chi_i(m)\chi_i(n)$.

Hints: Show $(m-1)(n-1) \equiv 0 \pmod 4$ and thus

5.P.6

$$\frac{mn-1}{2} \equiv \frac{m-1}{2} + \frac{n-1}{2} \pmod 2. \quad \text{Similarly } (m^*-1)(n^*-1) \equiv 0 \pmod{16} \Rightarrow$$

$$\frac{(mn)^*-1}{4} \equiv \frac{m^*-1}{4} + \frac{n^*-1}{4} \pmod 4, \quad \text{and use}$$

$$\frac{m^*n^*-1}{4} = \frac{(m^*n^*)^2-1}{8} = \frac{(mn)^2-1}{8} = \frac{(mn)^*-1}{4}\,.$$

Now use x_4 to show $m^*n^*=(mn)^*$.

b) Show that for odd r,s $\;x_4(m)^{\frac{rs-1}{2}} = x_4(m)^{\frac{r-1}{2}} x_4(m)^{\frac{s-1}{2}}$ (This is trivial,

but needed below.)

22. Suppose b is odd and $(a,b) = 1$. Define the extended Jacobi symbol (where

the denominator may be negative) by

$$\left(\frac{a}{b}\right) = \left(\frac{a}{|b|}\right)$$

where the symbol on the right is the symbol of §44.

Show

a) $\left(\frac{-1}{b}\right) = (-1)^{\frac{b-1}{2} + \frac{\operatorname{sgn} b-1}{2}}\,.$

    Hint: $\left(\frac{-1}{b}\right) = \left(\frac{-1}{|b|}\right) = x_4(|b|) = x_4(b\,\operatorname{sgn} b).$

b) $\left(\frac{2}{b}\right) = (-1)^{\frac{b^*-1}{4}} = (-1)^{\frac{b^2-1}{8}}\,.$

c) $\left(\frac{a}{b}\right) = (-1)^{\frac{a-1}{2}\frac{b-1}{2} + \frac{\operatorname{sgn} a -1}{2}\frac{\operatorname{sgn} b -1}{2}} \left(\frac{b}{a}\right)$

    Hint: Let $\alpha = \frac{\operatorname{sgn} a -1}{2}$ $\;\beta = \frac{\operatorname{sgn} b -1}{2}$ so that $a = (-1)^{\alpha}|a|$

$b = (-1)^{\beta}|b|$. Show first $\left(\frac{a}{b}\right) = \left(\frac{-1}{|b|}\right)^{\alpha}\left(\frac{-1}{|b|}\right)^{\beta}(-1)^{\frac{|a|-1}{2}\frac{|b|-1}{2}}\left(\frac{b}{a}\right)\,.$

Rewrite, after some manipulations involving 21a) for x_4

$$\left(\frac{a}{b}\right) = x_4(a)^{\frac{|b|-1}{2}}\, x_4(a)^{\frac{\operatorname{sgn} b-1}{2}}\, x_4(\operatorname{sgn} a)^{\frac{\operatorname{sgn} b-1}{2}}\left(\frac{b}{a}\right)$$

and use 21b).

Chapter 6
Binary Quadratic Forms

46. BASIC NOTIONS

An integral function is a sum of the form

$$F = \Sigma \; c_{\alpha_1 \ldots \alpha_n} x_1^{\alpha_1} \ldots x_n^{\alpha_n}$$

where $c_{\alpha_1 \ldots \alpha_n}$ is an integer constant and $\alpha_1, \ldots, \alpha_n$ are non negative integers.

An integral function is called <u>homogeneous</u> if the natural number $r = \alpha_1 + \alpha_2 + \ldots + \alpha_n$ is the same for each term in the sum. A homogeneous integral function with $\alpha_1 + \ldots + \alpha_n = r$ is called a form of degree r. When $r = 1$ then $F = c_1 x_1 + \ldots + c_n x_n$ is called a linear form. When $r = 2$,

$$F = c_{11} x_1^2 + c_{12} x_1 x_2 + \ldots + c_{1n} x_1 x_n + c_{21} x_2 x_1 + \ldots + c_{2n} x_2 x_n +$$
$$+ c_{31} x_3 x_1 + \ldots + c_{nn} x_n^2$$

is called a quadratic form. When $n = 2$ the quadratic form is called binary. We will concern ourselves in this chapter with such binary quadratic forms.

There are two basic problems for integral forms. Let $F(x,y,\ldots,w)$ be an integral form and m an integer. If the equation $F(x,y,\ldots,w) = m$ has an integer solution $(x_0,y_0,\ldots,w_0)$, then we call $(x_0,y_0,\ldots,w_0)$ a representation of m by F. We will also say that $F(x,y,\ldots,w)$ represents the integer m.

6.46.2

1st PROBLEM: Given an integral form $F(x,y,\ldots,w)$ and an integer m, we ask: is the equation $F(x,y,\ldots,w) = m$ solvable in integers?

2nd PROBLEM: If the equation $F(x,y,\ldots,w) = m$ is solvable in integers, find all the distinct representations $(x,y,\ldots,w)$ of m by F.

As examples we cite

I) The equation $x^2+y^2 = p$ where p is a prime number and $p \equiv 1 \pmod 4$ is solvable with integers x,y, and essentially in only one way (see §60).

II) Any positive integer can be represented as a sum of four squares of integers; that is, the integral quadratic form $F = x^2+y^2+z^2+t^2$ represents any positive integer. (We will not treat this problem, which does not involve a binary quadratic form).

The general expression of an integral binary quadratic form is

$$ax^2+bxy+cy^2 \tag{46.1}$$

where a,b,c are integers.

Since the equation

$$a'x^2+b'xy+c'y^2 = m'$$

is equivalent to

$$2a'x^2+2b'xy+2c'y^2 = 2m',$$

if we know how to solve each equation of the form

$$ax^2+2bxy+cy^2 = m \tag{46.2}$$

then we will know how to solve each equation of the form

$$ax^2+bxy+cy^2 = m. \tag{46.3}$$

6.46.3

Because of this, we will study the two problems introduced above for quadratic forms of the form (46.2) with the middle coefficient an even integer.

We will designate the quadratic form $ax^2+2bxy+cy^2$ by the symbol (a,b,c), and will call the integers a,b,c the first, second, third coefficients respectively. The variable x will be called the first variable and y the second variable.

47. AUXILIARY ALGEBRAIC FORMS

Let

$$f(x,y) = ax^2+2bxy+cy^2 \text{ where } a,b,c \text{ are arbitrary constants.}$$

The <u>polar form</u> of $f(x,y)$ is the bilinear (with respect to x,y and x',y')
form

$$f(x,y|x',y') = axx'+b(xy'+x'y)+cyy'$$
$$= (ax+by)x'+(bx+cy)y'$$
$$= (ax'+by')x+(bx'+cy')y.$$

As we see immediately, the polar form of $f(x,y)$ is symmetric with respect
to the pairs (x,y) and (x',y').

The determinant $-\begin{vmatrix} a & b \\ b & c \end{vmatrix} = \begin{vmatrix} b & a \\ c & b \end{vmatrix} = b^2 - ac$ is called the discriminant
of $f(x,y) = ax^2+2bxy+cy^2$ and will be designated by Δ.

We have the following easily verified identities; the fourth derived
from the first three,

$$af(x,y) = a^2x^2+2abxy+acy^2 = (ax+by)^2-\Delta y^2,$$
$$cf(x,y) = (bx+cy)^2-\Delta x^2,$$
$$bf(x,y) = (ax+by)(bx+cy)+\Delta xy,$$
$$f(x,y)f(x',y') = af(x,y)x'^2+2bf(x,y)x'y'+cf(x,y)y'^2$$
$$= (ax+by)^2x'^2+2(ax+by)(bx+cy)x'y'+(bx+cy)^2y'^2$$
$$-\Delta y^2x'^2+2\Delta xyx'y'-\Delta x^2y'^2$$
$$= \{(ax+by)x'+(bx+cy)y'\}^2-\Delta(x'y-xy')^2.$$

Hence

$$[f(x,y|x',y')]^2-f(x,y)f(x',y') = \Delta(xy'-x'y)^2. \tag{47.1}$$

48. LINEAR TRANSFORMATION OF THE QUADRATIC FORM $ax^2+2bxy+cy^2$

We set

$$x = \alpha x' + \beta y' \qquad \text{with } D = \begin{vmatrix} \alpha & \beta \\ \gamma & \delta \end{vmatrix} \neq 0. \qquad (48.1)$$
$$y = \gamma x' + \delta y'$$

Substituting these values into $f(x,y) = ax^2+2bxy+cy^2$,

we find

$$ax^2 + 2bxy + cy^2 = a(\alpha x'+\beta y')^2 + 2b(\alpha x'+\beta y')(\gamma x'+\delta y')+c(\gamma x'+\delta y')^2$$
$$= a'x'^2 + 2b'x'y' + c'y'^2,$$

where

$$a' = a\alpha^2 + 2b\alpha\gamma + c\gamma^2 = f(\alpha,\gamma),$$
$$b' = a\alpha\beta + b(\alpha\delta+\beta\gamma) + c\gamma\delta = f(\alpha,\gamma \mid \beta,\delta), \qquad (48.2)$$
$$c' = a\beta^2 + 2b\beta\delta + c\delta^2 = f(\beta,\delta).$$

To find the relation between the discriminant Δ' of the transformed

quadratic form $a'x'^2+2b'x'y'+c'y'^2$ and the discriminant Δ of

$ax^2+2bxy+cy^2$, we use the identity (47.1) setting $x = \alpha$, $y = \gamma$, $x' = \beta$,

$y' = \delta$. We find

$$[f(\alpha,\gamma \mid \beta,\delta)]^2 - f(\alpha,\gamma)f(\beta,\delta) = \Delta(\alpha\delta-\beta\gamma)^2$$

so, from (48.2)

$$b'^2 - a'c' = \Delta D^2,$$

that is

$$\Delta' = \Delta D^2. \qquad (48.3)$$

The above relations will be put in matrix form in the next section.

6.49.1

49. SUBSTITUTIONS AND COMPUTATIONS WITH THEM

The transition from the pair (x', y') to the pair $x = \alpha x' + \beta y'$, $y = \gamma x' + \delta y'$ we will call a substitution (or transformation) and we will write it in matrix form

$$S = \begin{pmatrix} \alpha & \beta \\ \gamma & \delta \end{pmatrix}, \qquad \begin{pmatrix} x \\ y \end{pmatrix} = \begin{pmatrix} \alpha & \beta \\ \gamma & \delta \end{pmatrix}\begin{pmatrix} x' \\ y' \end{pmatrix} = S\begin{pmatrix} x' \\ y' \end{pmatrix}, \tag{49.1}$$

(reading: from x', y' we go over to x, y by means of S), and we symbolize the corresponding transformation of the quadratic form (a,b,c) into (a', b', c') by

$$(a,b,c)S = (a', b', c') \text{ or } fS = f', \tag{49.2}$$

(reading: the quadratic form (a,b,c) is transformed into the quadratic form (a', b', c') by S).

We note that

$$\begin{pmatrix} x \\ y \end{pmatrix} = \begin{pmatrix} \alpha & \beta \\ \gamma & \delta \end{pmatrix}\begin{pmatrix} x' \\ y' \end{pmatrix} = \begin{pmatrix} \alpha x' + \beta y' \\ \gamma x' + \delta y' \end{pmatrix}. \tag{49.3}$$

For brevity we sometimes designate (x,y) and (x', y') by ξ and ξ' respectively, and (49.1) may be written more simply as $\xi = S\xi'$. We will use the same letter to designate a substitution and its matrix.

We assume that the determinant $D = \det S = \begin{vmatrix} \alpha & \beta \\ \gamma & \delta \end{vmatrix}$ of the substitution S is not 0. Then there exists an inverse substitution which we find by solving (48.1) for x' and y'. We find

$$\begin{aligned} x' &= \frac{\delta}{D} x - \frac{\beta}{D} y \\ y' &= -\frac{\gamma}{D} x + \frac{\alpha}{D} y \end{aligned} \qquad \text{with matrix } \begin{pmatrix} \frac{\delta}{D} & -\frac{\beta}{D} \\ -\frac{\gamma}{D} & \frac{\alpha}{D} \end{pmatrix} = \frac{1}{D}\begin{pmatrix} \delta & -\beta \\ -\gamma & \alpha \end{pmatrix}. \tag{49.4}$$

6.49.2

This substitution, and its matrix as well, we will designate by S^{-1}. Thus we have

$$\begin{pmatrix} x' \\ y' \end{pmatrix} = S^{-1} \begin{pmatrix} x \\ y \end{pmatrix}.$$

From the identity with respect to x', y',

$$a(\alpha x' + \beta y')^2 + 2b(\alpha x' + \beta y')(\gamma x' + \delta y') + c(\gamma x' + \delta y')^2 = a'x'^2 + 2b'x'y' + c'y'^2$$

we easily derive an identity with respect to x, y if we set

$$x' = \frac{\delta}{D} x - \frac{\beta}{D} y, \qquad y' = -\frac{\gamma}{D} x + \frac{\alpha}{D} y.$$ This new identity is none other than

$$ax^2 + 2bxy + cy^2 = a'\left(\frac{\delta}{D} x - \frac{\beta}{D} y\right)^2 + 2b'\left(\frac{\delta}{D} x - \frac{\beta}{D}\right)\left(-\frac{\gamma}{D} x + \frac{\alpha}{D} y\right)$$
$$+ c'\left(-\frac{\gamma}{D} x + \frac{\alpha}{D} y\right)^2,$$

which tells us that the quadratic form $a'x'^2 + 2b'x'y' + c'y'^2$ is transformed into $ax^2 + 2bxy + cy^2$ by S^{-1}:

$$f' S^{-1} = f, \qquad (a', b', c') S^{-1} = (a, b, c)$$

where $f' = a'x'^2 + 2b'x'y' + c'y'^2$ and $f = ax^2 + 2bxy + cy^2$.

We will now consider two substitutions

$$S: \begin{array}{l} x = \alpha x' + \beta y' \\ y = \gamma x' + \delta y' \end{array} \qquad \text{and} \qquad S_1: \begin{array}{l} x' = \alpha_1 x'' + \beta_1 y'' \\ y' = \gamma_1 x'' + \delta_1 y''. \end{array}$$

The composition SS_1 of the substitutions S and S_1 is the substitution which first takes x'', y'' into x', y' by means of S_1 and then takes x', y' into x, y by means of S:

$$x = \alpha(\alpha_1 x'' + \beta_1 y'') + \beta(\gamma_1 x'' + \delta_1 y'') = (\alpha\alpha_1 + \beta\gamma_1)x'' + (\alpha\beta_1 + \beta\delta_1)y''$$
$$y = \gamma(\alpha_1 x'' + \beta_1 y'') + \delta(\gamma_1 x'' + \delta_1 y'') = (\gamma\alpha_1 + \delta\gamma_1)x'' + (\gamma\beta_1 - \delta\delta_1)y''.$$

6.49.3

We notice that the matrix of the substitution SS_1 is equal to the product of the matrices of the substitutions S and S_1:

$$\begin{pmatrix} \alpha\alpha_1 + \beta\gamma_1, & \alpha\beta_1 + \beta\delta_1 \\ \gamma\alpha_1 + \delta\gamma_1, & \gamma\beta_1 + \delta\delta_1 \end{pmatrix} = \begin{pmatrix} \alpha & \beta \\ \gamma & \delta \end{pmatrix}\begin{pmatrix} \alpha_1 & \beta_1 \\ \gamma_1 & \delta_1 \end{pmatrix}.$$

Because of this the composition SS_1 is also called the product of the substitutions S and S_1 taken with the order: first S_1 then S. Attention to the order of the factors in the product is essential, because matrix multiplication does not satisfy the commutative law which the following example shows:

$$\begin{pmatrix} 0 & 1 \\ 1 & 2 \end{pmatrix}\begin{pmatrix} 1 & 0 \\ 1 & 1 \end{pmatrix} = \begin{pmatrix} 1 & 1 \\ 3 & 2 \end{pmatrix}.$$

$$\begin{pmatrix} 1 & 0 \\ 1 & 1 \end{pmatrix}\begin{pmatrix} 0 & 1 \\ 1 & 2 \end{pmatrix} = \begin{pmatrix} 0 & 1 \\ 1 & 3 \end{pmatrix}.$$

On the other hand, for three substitutions the associative law holds

$$(S_1 S_2)S_3 = S_1(S_2 S_3), \tag{49.5}$$

both for the substitutions themselves and for their matrices,

$$\left[\begin{pmatrix} \alpha_1 & \beta_1 \\ \gamma_1 & \delta_1 \end{pmatrix}\begin{pmatrix} \alpha_2 & \beta_2 \\ \gamma_2 & \delta_2 \end{pmatrix}\approx\begin{pmatrix} \alpha_3 & \beta_3 \\ \gamma_3 & \delta_3 \end{pmatrix}\right) = \begin{pmatrix} \alpha_1 & \beta_1 \\ \gamma_1 & \delta_1 \end{pmatrix}\left[\begin{pmatrix} \alpha_2 & \beta_2 \\ \gamma_2 & \delta_2 \end{pmatrix}\begin{pmatrix} \alpha_3 & \beta_3 \\ \gamma_3 & \delta_3 \end{pmatrix}\right]\approx,$$

as is easily verified by direct calculation. As is known, the

6.49.4

associative law permits the definition of products of three substitutions

$$S_1 S_2 S_3 = (S_1 S_2) S_3,$$

or, with induction, $n > 3$ substitutions

$$S_1 S_2 S_3 \ldots S_{n-1} S_n = (S_1 S_2 \ldots S_{n-1}) S_n.$$

By the associative law, the result depends only on the order of the factors but not on the positions of the parentheses, that is, the grouping of the factors. For example

$$S_1 S_2 S_3 S_4 S_5 = (S_1 S_2) S_3 (S_4 S_5) = S_1 (S_2 S_3 S_4) S_5.$$

In particular, we have to consider products $S_1 S_2 \ldots S_n$ with $S_1 = S_2 = \ldots = S_n = S$; we will designate this product by S^n where $n > 2$ is a natural number. We complete the symbolism by setting

$$S^1 = S, \quad S^0 = \text{(identity substitution) with matrix } \begin{pmatrix} 1 & 0 \\ 0 & 1 \end{pmatrix}$$

and

$$S^{-n} = (S^{-1})^n \qquad \text{for } n \text{ a natural number.}$$

It is easily proved that

$$S^m S^n = S^{m+n} \qquad \text{for } m, n \in \mathbf{Z}.$$

We note also that the determinant of the product of two substitutions is equal to the product of the determinants,

$$\det(S_1 S_2) = (\det S_1)(\det S_2), \tag{49.6}$$

as is easily verified by direct calculation.

We now introduce the concept of the transpose of a matrix in order to reformulate some of the results of §47 and §48 in matrix form. The transpose S^T of a matrix S is obtained by changing the rows of S into the

6.49.5

columns of S^T in the corresponding order. For example

$$\begin{pmatrix} \alpha & \beta \\ \gamma & \delta \end{pmatrix}^T = \begin{pmatrix} \alpha & \gamma \\ \beta & \delta \end{pmatrix}, \qquad \begin{pmatrix} x \\ y \end{pmatrix}^T = (x, y) \ .$$

The following law holds for an ℓ by m matrix A and an m by n matrix B:

$$(AB)^T = B^T A^T, \tag{49.7}$$

as is easily verified by direct calculation.

To the quadratic form $f(x,y) = ax^2 + 2bxy + cy^2$ we assign the matrix $Q = \begin{pmatrix} a & b \\ b & c \end{pmatrix}$. Designating the pair x,y (written vertically) by $\xi = \begin{pmatrix} x \\ y \end{pmatrix}$, we have

$$f(x,y) = \xi^T Q \xi = (x \ \ y) \begin{pmatrix} a & b \\ b & c \end{pmatrix} \begin{pmatrix} x \\ y \end{pmatrix}. \tag{49.8}$$

The substitution (48.1) may be written

$$\xi = S\xi', \qquad \begin{pmatrix} x \\ y \end{pmatrix} = \begin{pmatrix} \alpha & \beta \\ \gamma & \delta \end{pmatrix} \begin{pmatrix} x' \\ y' \end{pmatrix}. \tag{49.9}$$

Using (49.7) we have

$$\xi^T = \xi'^T S^T,$$

so that, introducing these into (49.8), we have

$$f'(x',y') = \xi'^T S^T Q S \xi'.$$

Hence, the matrix of f' is

$$S^T Q S = \begin{pmatrix} \alpha & \gamma \\ \beta & \delta \end{pmatrix} \begin{pmatrix} a & b \\ b & c \end{pmatrix} \begin{pmatrix} \alpha & \beta \\ \gamma & \delta \end{pmatrix} = \begin{pmatrix} a\alpha + b\gamma & b\alpha + c\gamma \\ a\beta + b\delta & b\beta + c\delta \end{pmatrix} \begin{pmatrix} \alpha & \beta \\ \gamma & \delta \end{pmatrix}$$

$$= \begin{pmatrix} f(\alpha,\gamma), & f(\alpha,\gamma \mid \beta,\delta) \\ f(\beta,\delta \mid \alpha,\gamma), & f(\beta,\delta) \end{pmatrix} = \begin{pmatrix} a' & b' \\ b' & c' \end{pmatrix} = Q', \tag{49.10}$$

which is the matrix form of (48.2).

Since, for a square matrix S, $\det S = \det S^T$, if we use (49.6) on this last matrix equation, we will have

$$(\det S^T)(\det Q)(\det S) = \det Q'$$

$$D(-\Delta)D \qquad\qquad = -\Delta'$$

$$\Delta D^2 = \Delta',$$

and we have a new derivation of (48.3).

6.49.6

The identity (47.1) is also a consequence of these relations, for in the equation (49.10) the equality

$$\begin{pmatrix} \alpha & \gamma \\ \beta & \delta \end{pmatrix}\begin{pmatrix} a & b \\ b & c \end{pmatrix}\begin{pmatrix} \alpha & \beta \\ \gamma & \delta \end{pmatrix} = \begin{pmatrix} f(\alpha,\gamma) & f(\alpha,\gamma|\beta,\delta) \\ f(\beta,\delta|\alpha,\gamma) & f(\beta,\delta) \end{pmatrix}$$

does not depend on $\begin{pmatrix} \alpha & \beta \\ \gamma & \delta \end{pmatrix}$ being the matrix of a substitution. We may replace α, β, γ, δ by x, x', y, y' respectively;

$$\begin{pmatrix} x & y \\ x' & y' \end{pmatrix}\begin{pmatrix} a & b \\ b & c \end{pmatrix}\begin{pmatrix} x & x' \\ y & y' \end{pmatrix} = \begin{pmatrix} f(x,y) & f(x,y|x',y') \\ f(x',y'|x,y) & f(x',y') \end{pmatrix},$$

from which, by (49.6), we obtain (47.1):

$$-\Delta(xy'-x'y)^2 = f(x,y)f(x',y')-[f(x,y|x',y')]^2.$$

50. UNIMODULAR TRANSFORMATIONS (OR UNIMODULAR SUBSTITUTIONS)

Let S be a substitution with matrix $S = \begin{pmatrix} \alpha & \beta \\ \gamma & \delta \end{pmatrix}$ where α, β, γ, δ are integers and $\alpha\delta-\beta\gamma \neq 0$. We represent the inverse substitution by $S^{-1} = \begin{pmatrix} \alpha' & \beta' \\ \gamma' & \delta' \end{pmatrix}$; we will then have $SS^{-1} = \begin{pmatrix} 1 & 0 \\ 0 & 1 \end{pmatrix}$. Let us assume that α', β', γ', δ' are (along with α, β, γ, δ) integers. The values of the determinants $\begin{vmatrix} \alpha & \beta \\ \gamma & \delta \end{vmatrix}$ and $\begin{vmatrix} \alpha' & \beta' \\ \gamma' & \delta' \end{vmatrix}$ will then be integers, and their product will be $\begin{vmatrix} 1 & 0 \\ 0 & 1 \end{vmatrix} = 1$. Hence the values of the determinants will be ± 1. We have shown that if α', β', γ', δ' are integers, then $\begin{vmatrix} \alpha & \beta \\ \gamma & \delta \end{vmatrix} = \alpha\delta-\beta\gamma = \pm 1$. Conversely, if $\begin{vmatrix} \alpha & \beta \\ \gamma & \delta \end{vmatrix} = \pm 1$ then α', β', γ', δ' are integers, because, (as is easily verified by calculation)

$$S^{-1} = \begin{pmatrix} \alpha' & \beta' \\ \gamma' & \delta' \end{pmatrix} = \begin{pmatrix} \dfrac{\delta}{\pm 1} & \dfrac{-\beta}{\pm 1} \\ \dfrac{-\gamma}{\pm 1} & \dfrac{\alpha}{\pm 1} \end{pmatrix} .$$

Below we will concern ourselves only with transformations whose matrices consist of integers satisfying the condition

$$\begin{vmatrix} \alpha & \beta \\ \gamma & \delta \end{vmatrix} = +1.$$

We will call these unimodular transformations or unimodular substitutions. If $S = \begin{pmatrix} \alpha & \beta \\ \gamma & \delta \end{pmatrix}$ is unimodular, then the same is true of the inverse transformation $S^{-1} = \begin{pmatrix} \delta & -\beta \\ -\gamma & \alpha \end{pmatrix}$. Thus the set of unimodular transformations forms a group with respect to the operation of composition (or multiplication) of two transformations.

6.50.2

Let now $\binom{x}{y} = S\binom{x'}{y'}$, where S is a unimodular transformation. Then to a pair (x',y') of integers corresponds a pair (x,y) which also consists of integers. Conversely, to a pair (x,y) of integers corresponds a pair (x',y') of integers.

Two unimodular substitutions $\begin{pmatrix} \alpha & \beta \\ \gamma & \delta \end{pmatrix}$ and $\begin{pmatrix} \alpha & \beta_1 \\ \gamma & \delta_1 \end{pmatrix}$ are called <u>parallel</u>. We pose the problem:

<u>PROBLEM</u> Given an $S_0 = \begin{pmatrix} \alpha & \beta_0 \\ \gamma & \delta_0 \end{pmatrix}$, determine all the substitutions parallel to it.

<u>SOLUTION</u>. Let $S = \begin{pmatrix} \alpha & \beta \\ \gamma & \delta \end{pmatrix}$ be parallel to $S_0 = \begin{pmatrix} \alpha & \beta_0 \\ \gamma & \delta_0 \end{pmatrix}$. Then we will have

$$S_0^{-1}S = \begin{pmatrix} \delta_0 & -\beta_0 \\ -\gamma & \alpha \end{pmatrix}\begin{pmatrix} \alpha & \beta \\ \gamma & \delta \end{pmatrix} = \begin{pmatrix} \delta_0\alpha - \beta_0\gamma & \delta_0\beta - \beta_0\delta \\ -\gamma\alpha + \alpha\gamma & -\gamma\beta + \alpha\delta \end{pmatrix}$$

$$= \begin{pmatrix} 1 & \delta_0\beta - \beta_0\delta \\ 0 & 1 \end{pmatrix} = \begin{pmatrix} 1 & k \\ 0 & 1 \end{pmatrix},$$

where k is an integer. Thus

$$S = S_0 S_0^{-1} S = \begin{pmatrix} \alpha & \beta_0 \\ \gamma & \delta_0 \end{pmatrix}\begin{pmatrix} 1 & k \\ 0 & 1 \end{pmatrix} = \begin{pmatrix} \alpha & \beta_0 + k\alpha \\ \gamma & \delta_0 + k\gamma \end{pmatrix}.$$

On the other hand, a substitution with matrix $\begin{pmatrix} \alpha & \beta_0 + k\alpha \\ \gamma & \delta_0 + k\gamma \end{pmatrix}$ with $k \in Z$ is a unimodular substitution parallel to $\begin{pmatrix} \alpha & \beta_0 \\ \gamma & \delta_0 \end{pmatrix}$. Thus we get all the substitutions parallel to the substitution $\begin{pmatrix} \alpha & \beta_0 \\ \gamma & \delta_0 \end{pmatrix}$ if we make k in $\begin{pmatrix} \alpha & \beta_0 + k\alpha \\ \gamma & \delta_0 + k\gamma \end{pmatrix}$ run through the set of Z of integers.

6.51.1

51. EQUIVALENCE OF QUADRATIC FORMS

Let $f = (a,b,c)$, $f' = (a',b',c')$, and

$fS = f'$.

(We remind ourselves that all the substitutions under consideration are unimodular). Then f is said to be _equivalent_ to f' and we write

$f \sim f'$.

PROPERTIES The relation defined above on the set of integral quadratic

forms is an equivalence relation, for it has the three properties

I) $f \sim f$, since $f\begin{pmatrix} 1 & 0 \\ 0 & 1 \end{pmatrix} = f$.

II) $f \sim f' \Rightarrow f' \sim f$ since $fS = f' \Rightarrow f = f'S^{-1}$.

III) $f \sim f'$ and $f' \sim f'' \Rightarrow f \sim f''$

since $f' = fS_1$ and $f'' = f'S_2 \Rightarrow f'' = fS_1S_2$.

We now may conclude that the set of quadratic forms is divided into classes of equivalent quadratic forms. Two equivalence classes either coincide or have no common member. We will designate the equivalence class which contains the quadratic form $f_0 = (a_0, b_0, c_0)$ by $c\ell(f_0)$. Thus we have $c\ell(f) = c\ell(f')$ if and only if $f \sim f'$.

We present some other properties of equivalent quadratic forms:

IV) Let $f \sim f'$, Δ the discriminant of f and Δ' the

discriminant of f'. Then

$\Delta = \Delta'$.

6.51.2

In fact, let $fS = f'$ with $S = \begin{pmatrix} \alpha & \beta \\ \gamma & \delta \end{pmatrix}$, where $D = \begin{vmatrix} \alpha & \beta \\ \gamma & \delta \end{vmatrix}$. As we saw in §48,

$$\Delta' = \Delta D^2 = \Delta \, 1^2 = \Delta.$$

V) We designate by kf, with integer k, the form $kf = (ka,kb,kc) = kax^2 + 2kbxy + kcy^2$. If $f \sim f'$ then $kf \sim kf'$.

In fact, if $f\begin{pmatrix} \alpha & \beta \\ \gamma & \delta \end{pmatrix} = f'$ then $kf\begin{pmatrix} \alpha & \beta \\ \gamma & \delta \end{pmatrix} = kf'$.

VI) If $f \sim f'$ then the forms f and f' represent the same integers; in other words, if the equation $f(x,y) = m$, where m is an integer, is solvable for integers x,y, then $f'(x',y') = m$ will be solvable with integers x',y'.

Proof Let $fS = S'$, where S: $\begin{aligned} x &= \alpha x' + \beta y' \\ y &= \gamma x' + \delta y' \end{aligned}$. We have the identity with respect to x',y':

$$f(\alpha x' + \beta y', \ \gamma x' + \delta y') = f'(x',y'). \tag{51.1}$$

Let now $f(x_0,y_0) = m$; that is, (x_0,y_0) is a representation of the integer m by the quadratic form f. To the pair (x_0,y_0) we correspond by the inverse substitution S^{-1} the pair

$$x_0' = \delta x_0 - \beta y_0, \qquad y_0' = -\gamma x_0 + \alpha y_0. \tag{51.2}$$

We introduce these values into the identity (51.1); we find

$$\begin{aligned}
f'(x_0',y_0') &= f(\alpha x_0' + \beta y_0', \ \gamma x_0' + \delta y_0') \\
&= f(\alpha(\delta x_0 - \beta y_0) + \beta(-\gamma x_0 + \alpha y_0), \ \gamma(\delta x_0 - \beta y_0) + \delta(-\gamma x_0 + \alpha y_0)) \\
&= f((\alpha\delta - \beta\gamma)x_0, \ (\alpha\delta - \beta\gamma)y_0) = f(x_0, y_0) = m.
\end{aligned}$$

Hence the integer m is also represented by the quadratic form f'; the representing pair x_0', y_0' is given by (51.2) and corresponds to the original representing pair x_0, y_0 by S^{-1}, the substitution which transforms the quadratic form f' into the equivalent form f.

6.52.1

52. SUBSTITUTIONS PARALLEL TO $\begin{pmatrix} 0 & -1 \\ 1 & 0 \end{pmatrix}$

Later we will need the substitutions which are parallel to $T = \begin{pmatrix} 0 & -1 \\ 1 & 0 \end{pmatrix}$. These are, according to §50, the following:

$$T_k = \begin{pmatrix} 0 & -1+0 \; k \\ 1 & 0+1 \; k \end{pmatrix} = \begin{pmatrix} 0 & -1 \\ 1 & k \end{pmatrix} \qquad \text{with } k \in \mathbf{Z}.$$

We consider now a quadratic form $f = (a,b,a')$. We will transform it by T_k, setting

$$x = 0 \; x'-1 \; y' = -y'$$
$$y = 1 \; x'+k \; y' = x'+ky'.$$

We find

$$fT_k = a(-y')^2 + 2b(-y')(x'+ky')+a'(x'+ky')^2$$
$$= a'x'^2+2(-b+a'k)x'y'+(a-2bk+a'k^2)y'^2$$
$$= (a',b',a''),$$

where

$$b' = -b+a'k \quad \text{and} \quad a'' = a-2bk+a'k^2.$$

The quadratic form (a',b',a'') is called right adjacent to (a,b,a'); and (a,b,a') is called left adjacent to (a',b',a'').

For $k = 0$ the right adjacent form to (a,b,a') is

$$fT_0 = (a',-b,a).$$

Naturally $(a,b,a') \sim (a',b',a'')$ so, in particular, $(a,b,a') \sim (a',-b,a)$. From the relation $b' = -b+a'k$ with $k \in \mathbf{Z}$, it follows that, in the case $a' \neq 0$, the second coefficients of all of the forms right adjacent to (a,b,a') belong to a well defined residue class mod $|a'|$, the class containing $-b$. On the other hand, let an integer $b' \equiv -b \pmod{|a'|}$ be given. Then there exists a unique form with second coefficient b' which

6.52.2

is right adjacent to (a,b,a'). Indeed, it suffices to take k such that $b'+b = ka'$, so k is equal to the integer $\dfrac{b'+b}{a'}$. Then

$$-b+a'k = -b+a' \dfrac{b'+b}{a'} = b' \text{ and we will have}$$

$$(a,b,a')\begin{pmatrix} 0 & -1 \\ 1 & k \end{pmatrix} = (a',b',a'') \ .$$

53. REDUCTION OF THE FIRST BASIC PROBLEM OF §46

Let $f = (a,b,c)$ and $m \neq 0$ an integer. Let (x_0,y_0) be a representation of m by f, that is

$$ax_0^2 + 2bx_0y_0 + cy_0^2 = m. \tag{53.1}$$

The integers x_0,y_0 cannot both vanish (since $m \neq 0$), so the greatest common divisor $(x_0,y_0) = \delta > 0$ exists and we have $x_0 = x_1\delta$, $y_0 = y_1\delta$, and $(x_1,y_1) = 1$. Then (53.1) may be written

$$\delta^2(ax_1^2 + 2bx_1y_1 + cy_1^2) = m,$$

from which we conclude that δ^2 is a divisor of m; $m \equiv 0 \pmod{\delta^2}$. Then we have the equation

$$ax_1^2 + 2bx_1y_1 + cy_1^2 = \frac{m}{\delta^2}.$$

Hence, if m is representable by f and if δ is the greatest common divisor of the pair x_0,y_0 which represent m, then δ^2 divides m and the integer m/δ^2 is also representable by f, with a pair x_1,y_1 for which $(x_1, y_1) = 1$.

A representation is called primitive if the greatest common divisor $(x,y) = 1$. The problem of finding the representations of an integer m by a quadratic form $f = (a,b,c)$ then reduces to the following: Find first all the primitive representations of m/δ^2, if they exist, where δ runs through all positive integers whose squares divide m. Second, from these primitive representations of m/δ^2, form, by multiplication by δ, all the representations of m. From the above it follows that if m has no divisors which are squares other than $1^2 = 1$, then there are no imprimitive representations of m.

6.53.2

We have now first of all to solve the following problem: Given a quadratic form f and an integer m ≠ 0, find all the primitive representations of m by f, if any exist.

Assume now that α, γ is a primitive representation of m by f, that is

$$f(\alpha,\gamma) = a\alpha^2 + 2b\alpha\gamma + c\gamma^2 = m \text{ with } (\alpha, \gamma) = 1.$$

We ask; what may we conclude from this assumption? Since $(\alpha, \gamma) = 1$, the Diophantine equation $\alpha x - \gamma y = 1$ has solutions and, in fact, we know (§24) that, if $x = \delta_0$, $y = \beta_0$ is one solution then all solutions have the form

$$x = \delta_0 + k\gamma \qquad y = \beta_0 + k\alpha \qquad \text{with } k \in \mathbb{Z}.$$

We consider the substitution

$$S_k = \begin{pmatrix} \alpha & \beta_0 + k\alpha \\ \gamma & \delta_0 + k\gamma \end{pmatrix}$$

which has determinant $\alpha\delta_0 - \gamma\beta_0 = 1$. We form the transformed quadratic form (see §48)

$$fS_k = (f(\alpha,\gamma),\ f(\alpha,\gamma|\beta_0 + k\alpha, \delta_0 + k\gamma),\ f(\beta_0 + k\alpha,\ \delta_0 + k\gamma))\ .$$

Because of the bilinearity of $f(\alpha,\gamma|\beta,\delta)$ we have

$$f(\alpha,\gamma|\beta_0 + k\alpha,\ \delta_0 + k\gamma) = f(\alpha,\gamma|\beta_0,\ \delta_0) + f(\alpha,\gamma|\ k\alpha,\ k\gamma)$$

so

$$fS_k = (m,\ n_0 + kf(\alpha,\gamma|\alpha,\gamma),\ \ell)$$

where

$$n_0 = f(\alpha,\gamma|\beta_0,\delta_0) \quad \text{and} \quad \ell = f(\beta_0 + k\alpha,\ \delta_0 + k\gamma).$$

However, we have

$$f(\alpha,\gamma|\alpha,\gamma) = f(\alpha,\gamma) = m.$$

Consequently

$$fS_k = (m,\ n_0+km,\ \ell).$$

Since the discriminant of fS_k equals the discriminant of f (§48),

$$(n_0+ km)^2- m\ell = b^2- ac = \Delta,$$

and thus

$$(n_0+ km)^2 \equiv \Delta \pmod{|m|}.$$

Hence, if $(\alpha,\ \gamma)$ is a primitive representation of m by f, then there exist infinitely many unimodular substitutions $S = \begin{pmatrix} \alpha & \beta \\ \gamma & \delta \end{pmatrix}$ such that

$$fS = (m,\ n,\ \ell);$$

so $f \sim (m,\ n,\ \ell)$, where the second coefficients n all belong to a definite residue class mod $|m|$ from among the classes which satisfy the congruence $x^2 \equiv \Delta \pmod{|m|}$.

Conversely, suppose $f = (a,\ b,\ c)$ is equivalent to a quadratic form $(m,\ n,\ \ell)$ where $n^2 \equiv \Delta \pmod{|m|}$, and let $S = \begin{pmatrix} \alpha & \beta \\ \gamma & \delta \end{pmatrix}$ be a unimodular substitution which transforms f into $(m,\ n,\ \ell)$; then $f(\alpha,\gamma) = m$ and the pair α,γ is a primitive representation of m by f.

From the above we conclude the following:

To find the primitive representations of an integer m by $f = (a,\ b,\ c)$ we first determine a complete system of solutions of the congruence

$$x^2 \equiv \Delta \pmod{|m|}.$$

6.53.4

Let n_j $(j = 1,\ldots,k)$ be these solutions. We form the quadratic form

$$\left(m, \ n_j, \ \frac{n_j^2 - \Delta}{m}\right)$$

and must answer the following two questions:

1) Is it true that $(a, b, c) \sim \left(m, n_j, \dfrac{n_j^2 - \Delta}{m}\right)$?

2) If it is true that $(a, b, c) \sim \left(m, n_j, \dfrac{n_j^2 - \Delta}{m}\right)$, then

   what are the distinct unimodular substitutions which transform

   (a, b, c) into $\left(m, n_j, \dfrac{n_j^2 - \Delta}{m}\right)$?

If we determine all these substitutions $\left(\begin{smallmatrix} \alpha & \beta \\ \gamma & \delta \end{smallmatrix}\right)$ then the pairs α, γ will be

all the primitive representations of m by $f = (a, b, c)$.

Thus we have reduced the two basic problems of §46 for binary

quadratic forms to the solution of the following two problems:

 I) Given two quadratic forms f and ϕ, is it true that $f \sim \phi$?

II) Let $f \sim \phi$. Determine all unimodular substitutions S for which $fS = \phi$.

54. REDUCED QUADRATIC FORMS WITH DISCRIMINANT $\Delta < 0$

Let $f = (a,b,c)$ with $\Delta = b^2-ac < 0$. Set $D = -\Delta > 0$*. We have $ac > b^2 \geq 0$, so the integers a,c are non-zero and have the same sign. According to §47

$$af = (ax+by)^2 - \Delta y^2 = (ax+by)^2 + Dy^2.$$

Thus $af \geq 0$ and if $af(x,y) = 0$ then, first, $y = 0$ and this will then imply $x = 0$. Thus $af = 0$ if and only if $x = y = 0$. Consequently, apart from 0, f may represent only positive integers if $a > 0$ and only negative integers if $a < 0$. The quadratic forms with discriminant $\Delta < 0$ are divided then into two categories: the category with $a > 0$ and the category with $a < 0$. We will call those of the first category positive forms because they represent only positive integers when $|x|+|y| \neq 0$. Those of the second category are called negative forms, and represent only negative integers when $|x| + |y| \neq 0$. A positive form cannot be equivalent to a negative form because they do not represent the same integers.

If f and φ are both negative forms and $f \sim \varphi$, then $-f$ and $-\varphi$ are both positive forms and $-f \sim -\varphi$. Thus the study of negative forms may be reduced to the study of positive forms. We now turn to the study of positive forms.

Let f and φ be two positive forms with the same discriminant. A consequence of this hypothesis is that the first coefficient of each

* Since all substitutions are now unimodular, we put the letter D to a new use.

6.54.2

form, as well as the third, is a positive integer. We pose the question:
is $f \sim \varphi$? To provide an answer we introduce the following definition:

<u>DEFINITION</u> A form $f = (a,b,c)$ with negative discriminant is called
reduced if
$$2|b| \leqslant a \leqslant c$$
and additionally, if any of the inequalities reduces to an
equality, then $b \geqslant 0$.

For example, $2x^2+2xy+3y^2$ is reduced, since $2|1| = 2 \leqslant 3$ and $b = 1 \geqslant 0$;
$2x^2+3y^2$ is also reduced. But $2x^2-2xy+3y^2$ is not reduced, since it
violates the additional condition.

From the definition it follows that in a reduced quadratic form the
second coefficient belongs to a system of least absolute residues mod a.

<u>PROPOSITION 1</u> Every positive quadratic form is equivalent to a reduced
form

<u>Proof</u> Let $f = (a,b,a')$ be a positive non-reduced form. Recall that
$a,a' > 0$. We consider the result of transforming the original form
by $T_k = \begin{pmatrix} 0 & -1 \\ 1 & k \end{pmatrix}$:
$$(a,b,a')T_k = (a',b',a'')$$
and select the integer k so that b', which is congruent to $-b \pmod{a'}$,
will belong to a system of absolute least residues mod a'.
More precisely
$$-\frac{a'}{2} < b' \leqslant \frac{a'}{2} \tag{54.1}$$
and thus $2|b'| \leqslant a'$. We now have four possibilities 1) $a' > a''$
2) $2|b'| = a' \leqslant a''$ and $b' < 0$ 3) $2|b'| < a' = a''$ and $b' < 0$
4) (a',b',a'') is reduced.

6.54.3

In the first case we repeat the previous procedure, determining k' so that

$$(a', b', a'')T_{k'} = (a'', b'', a'''),$$

where $b'' \equiv -b' \pmod{a''}$ and $2|b''| < a''$. If we again have case 1: $a'' > a'''$, we again repeat the procedure

$$(a'', b'', a''')T_{k''} = (a''', b''', a^{IV}).$$

We obtain a descending sequence $a' > a'' > a''' > \ldots$ of natural numbers. Thus the procedure must terminate, let us say at the $(r-1)^{st}$ step:

$$(a^{(r-1)}, b^{(r-1)}, a^{(r)})T_{k^{(r-1)}} = (a^{(r)}, b^{(r)}, a^{(r+1)})$$

with $2|b^{(r)}| < a^{(r)} < a^{(r+1)}$.

At this point, one of cases 2, 3 or 4 must hold. Case 2 cannot hold however, because we would have $2|b^{(r)}| = a^{(r)}$, so $a^{(r)}$ would be even, and, because of the way the least absolute residues were chosen, (54.1), we would have $b^{(r)} = \dfrac{a^{(r)}}{2} > 0$. In case 3, we have $2|b^{(r)}| < a^{(r)} = a^{(r+1)}$ and $b^{(r)} < 0$. We use the substitution T_0 and obtain

$$(a^{(r)}, b^{(r)}, a^{(r+1)})T_0 = (a^{(r+1)}, -b^{(r)}, a^{(r)}),$$

which is reduced. We have proved that any positive quadratic form is equivalent to a reduced quadratic form, which may be determined by subjecting a given non-reduced positive quadratic form to a finite series of substitutions of the form $\begin{pmatrix} 0 & -1 \\ 1 & k \end{pmatrix}$.

<u>PROPOSITION 2</u> If $f = (a,b,c)$ is positive and reduced, then a is the smallest positive integer which may be represented by f.

Proof Since $2|b| \leqslant a \leqslant c$, we may set

$$a = 2|b|+p \qquad c = 2|b|+p+q \qquad \text{with } p \geqslant 0, q \geqslant 0. \qquad (54.2)$$

We have

$$f = (2|b|+p)x^2+2bxy+(2|b|+p+q)y^2$$

$$= |b|(2x^2 \pm 2xy+2y^2) + p(x^2+y^2) + qy^2.$$

The quadratic form $2x^2 \pm 2xy+2y^2 = (2,\pm 1,2)$ is positive since its

discriminant is $1-4 = -3 < 0$, and its first coefficient $2 > 0$. Thus

it represents only positive integers for $|x|+|y| \neq 0$.

We have, for $x = 1, y = 0$

$$f(1,0) = 2|b|+p = a$$

and

$$f(x,y)-a = 2|b|(x^2 \pm 2xy+y^2-1) + p(x^2+y^2-1) + qy^2, \qquad (54.3)$$

where the three terms on the right are non-negative for integers x,y

satisfying $|x|+|y| \neq 0$. This proves the proposition.

PROPOSITION 3 If $f = (a,b,c)$ is positive and reduced and $a < c$ then the

integer a (the least positive integer representable by f) is

represented only by the pairs $x = \pm 1, y = 0$.

Proof Since the hypothesis $a > c$ implies that $q \geqslant 1$ in the relations

(54.2), we will have, according to (54.3), $f(x,y)-a = 0$ only if $y =$

0. Then $f(x,y) = a$ becomes $ax^2 = a$ which implies $x = \pm 1$ since $a \neq 0$.

PROPOSITION 4 Two reduced quadratic forms (a,b,c) and $(a´,b´,c´)$ are

equivalent only if they coincide; that is

$$(a,b,c) \sim (a´,b´,c´) \Rightarrow a = a´, b = b´, c = c´.$$

Proof Since two equivalent quadratic forms represent the same

numbers, we have

a = (least positive integer representable by (a,b,c))

= (least positive integer representable by (a´,b´,c´))

= a´,

that is, a = a´. We distinguish two cases,

I) a < c. Then, since by the hypothesis

$$fS = (a,b,c)\begin{pmatrix} \alpha & \beta \\ \gamma & \delta \end{pmatrix} = (a´,b´,c´) \qquad \text{with } \alpha\delta-\beta\gamma = 1,$$

we will have (§48)

$$f(\alpha,\gamma) = a´ = a.$$

So, by the preceeding proposition, we have $\alpha = \pm 1$, $\gamma = 0$;

$$\begin{pmatrix} \alpha & \beta \\ \gamma & \delta \end{pmatrix} = \begin{pmatrix} \pm 1 & \beta \\ 0 & \delta \end{pmatrix}.$$

But then $\alpha\delta-\beta\gamma = 1$ implies

$$\begin{pmatrix} \alpha & \beta \\ \gamma & \delta \end{pmatrix} = \begin{pmatrix} \pm 1 & \beta \\ 0 & \pm 1 \end{pmatrix};$$

that is $\alpha = 1$, $\delta = 1$ or $\alpha = -1$, $\delta = -1$.

As we know (§48), we have also

$$b´ = f(\alpha,\gamma|\beta,\delta) = a\alpha\beta+b(\alpha\delta+\beta\gamma)+c\gamma\delta = \pm a\beta+b;$$

so

$$b \equiv b´ \qquad (\text{mod } a).$$

Since b and b´ are both least positive residues mod a = a´, we

have b = b´. Then from b´ = ±aβ+b it follows that β = 0. Thus

the substitution $\begin{pmatrix} \alpha & \beta \\ \gamma & \delta \end{pmatrix}$ is simply $\begin{pmatrix} \pm 1 & 0 \\ 0 & \pm 1 \end{pmatrix}$ and thus

$(a´,b´,c´) = (a\ b\ c)\begin{pmatrix} \pm 1 & 0 \\ 0 & \pm 1 \end{pmatrix} = (a\ b\ c)$ as required.

II) a = c. Then we will have a´ = c´, for if this were not true,

then we would have a´ < c´, and, applying case I to

$(a´,b´,c´) \sim (a,b,c)$ we would have a = a´ < c´ = c, a

6.54.6

contradiction. So we have $a = a' = c = c'$ and $b > 0$,
$b' > 0$. Moreover, we have $b^2-a^2 = \Delta = \Delta' = b'^2-a'^2$ so $b^2 = b'^2$.
Since $b, b' > 0$, we have $b = b'$, so $(a,b,c) = (a',b',c')$.

We return now to the problem: Given two positive quadratic forms f
and ϕ, is it true that $f \sim \phi$?

The answer is immediate from what we have shown above: find the
reduced forms F and Φ which are equivalent to f and ϕ. Then $f \sim \phi$ if and
only if $F = \Phi$.

In case $f \sim \phi$, we have $F = \Phi$ and

$$fS = F \qquad \phi S_1 = \Phi,$$

where S and S_1 are unimodular substitutions. Thus

$$\phi = \Phi S_1^{-1} = F S_1^{-1} = f S S_1^{-1}.$$

Hence we will transform f into ϕ when we apply the unimodular
transformation SS_1^{-1}.

It is now possible to answer the first problem of §46 when
$f = (a,b,c)$ has negative discriminant; Given an integer m and a positive
quadratic form $f = (a,b,c)$, is the equation $m = f(x,y)$ solvable? We
delay further discussion of the problem, however, until we can also solve
the second problem, for which we need the material in §57.

55. THE NUMBER OF CLASSES OF EQUIVALENT FORMS WITH DISCRIMINANT $\Delta < 0$

We consider the set of quadratic forms with discriminant $\Delta < 0$. In this set there is a relation of equivalence between the elements, so the set is decomposed into equivalence classes of quadratic forms. We will show that for each given Δ the number of these classes is finite. We will call this number the class number of the discriminant Δ.

As we have seen, each quadratic form with discriminant Δ is equivalent to a uniquely determined reduced form with the same discriminant Δ. Hence the equivalence classes of quadratic forms with discriminant Δ are in one to one correspondence with the reduced quadratic forms of discriminant Δ.

Now let (a,b,c) be a reduced quadratic form with discriminant $\Delta = -D < 0$. According to the definition of reduced quadratic forms, we have $2|b| \leqslant a \leqslant c$, and thus

$$4b^2 \leqslant a^2 \leqslant ac = b^2 + D.$$

Hence

$$3b^2 \leqslant D \quad \text{and} \quad |b| \leqslant \sqrt{D/3},$$

so the values which b may assume are the following

$$b = 0, \pm 1, \pm 2, \ldots, \pm \left[\sqrt{D/3}\right].$$

With each value of b are connected only a finite number of coefficients a and c, because $ac = b^2 + D$. To find them we have only to decompose $b^2 + D$ into products of two positive integer factors. Thus it has been proved that the number of reduced quadratic forms with discriminant Δ, and hence also the number of classes of equivalent forms with discriminant Δ, is finite.

6.55.2

Below we determine the reduced quadratic forms with discriminant

$\Delta = -1,-2,\ldots,-13$; so $D = 1,2,\ldots,13$.

When $D = 1$ then $|b| < [\sqrt{1/3}]$, so $b = 0$, $ac = 1$, so $a = c = 1$.

When $D = 2$ then $|b| < [\sqrt{2/3}]$, so $b = 0$, $ac = 2$, so $a = 1$ $c = 2$.

When $D = 3$ then $|b| < [\sqrt{3/3}]$, so $b = 0$, $ac = 3$, $a = 1$, $c = 3$ or $b = \pm 1$,

   $ac = 4$, $a = 2$, $c = 2$ or $a = 1$, $c = 4$. Only $b = 1$ $a = 2$ $c = 2$ leads

   to a reduced form; $(2,-1,2)$, and $(1,\pm 1,4)$ are not reduced.

When $D = 4$ then $b = 0,\pm 1$. If $b = 0$ then $ac = 4$ and $a = 1$, $c = 4$ or

   $a = 2 = c$. If $b = \pm 1$ then $ac = 5$, $a = 1$ $c = 5$ but $(1,\pm 1,5)$ is not

   reduced.

When $D = 5$ then $b = 0,\pm 1$. If $b = 0$ then $ac = 5$, $a = 1$, $c = 5$. If $b =$

   ± 1, $ac = 6$, so $a = 2$, $c = 3$ or $a = 1$, $c = 6$. $(1,\pm 1,6)$ is not

   reduced, nor is $(2,-1,3)$ since $2|1| = 2$ which requires $b > 0$.

When $D = 6$ then $b = 0,\pm 1$. If $b = 0$ then $ac = 6$ and $a = 1$, $c = 6$ or

   $a = 2$, $c = 3$. If $b = \pm 1$ $ac = 7$ so $a = 1$ $c = 7$ but $(1,\pm 1,7)$ is not

   reduced.

When $D = 7$, $b = 0,\pm 1$. If $b = 0$ then $ac = 7$ so $a = 1$, $c = 7$.

   If $b = \pm 1$ then $ac = 8$ so $a = 1$, $c = 8$ or $a = 2$, $c = 4$. But $(1,\pm 1,8)$

   is not reduced, nor is $(2,-1,4)$ since $2|1| = 2$ requires $b > 0$.

When $D = 12$, $b = 0,\pm 1,\pm 2$. If $b = 0$ then $ac = 12$ so $a = 1$, $c = 12$ or

   $a = 2$, $c = 6$ or $a = 3$, $c = 4$. If $b = \pm 1$ then $ac = 13$, $a = 1$, $c = 13$

   but $(1,\pm 1,13)$ is not reduced. If $b = \pm 2$ then $ac = 16$, so $a = 1$,

   $c = 16$ or $a = 2$, $c = 8$, or $a = 4$, $c = 4$, But $(1,\pm 2,16)$, $(2,\pm 2,8)$ are

   not reduced, nor is $(4,-2,4)$.

6.55.3

In the same way we find the reduced quadratic forms with D = 8,9,10,11,13. We compile the information in the following table:

D	REDUCED FORMS	CLASS NUMBER
1	(1,0,1)	1
2	(1,0,2)	1
3	(1,0,3),(2,1,2)	2
4	(1,0,4), (2,0,2)	2
5	(1,0,5),(2,1,3)	2
6	(1,0,6),(2,0,3)	2
7	(1,0,7),(2,1,4)	2
8	(1,0,8),(2,0,4),(3,1,3)	3
9	(1,0,9),(3,0,3),(2,1,5)	3
10	(1,0,10),(2,0,5)	2
11	(1,0,11),(2,1,6),(3,1,4),(3,-1,4)	4
12	(1,0,12),(2,0,6),(3,0,4),(4,2,4)	4
13	(1,0,13),(2,1,7)	2

As is obvious, among the reduced forms corresponding to a given value of D is x^2+Dy^2; this is called the <u>principle</u> form with discriminant $\Delta = -D$.

56. THE ROOTS OF A QUADRATIC FORM

Let $f = (a,b,c) = ax^2+2bxy+cy^2$ with $a \neq 0$ and $\Delta = b^2-ac \neq 0$. The roots of f are the solutions of the equation $ax^2+2bx+c = 0$. When $\Delta < 0$ we set $w = i\sqrt{ac-b^2} = i\sqrt{-\Delta}$ and when $\Delta > 0$ we set $w = -\sqrt{b^2-ac} = -\sqrt{\Delta}$.* The roots of f are

$$\omega = \frac{-b+w}{a} \qquad \eta = \frac{-b-w}{a} ,$$

so $\omega+\eta = -2b/a$ and $\omega\eta = (b^2-w^2)/a^2 = c/a$. We call ω the first root and η the second root.

PROPOSITION 1 We have

$$f = a(x-\omega y)(x-\eta y).$$

PROOF In fact

$$a(x-\omega y)(x-\eta y) = a\left(x^2-(\omega+\eta)xy+\omega\eta y^2\right) = a\left(x^2-(-\frac{2b}{a})xy+\frac{c}{a}y^2\right)$$
$$= ax^2+2bxy+cy^2.$$

PROPOSITION 2 IF (a,b,c) and (a',b',c') have the same discriminant and $aa' \neq 0$ and $\omega = \omega'$, $\eta = \eta'$ then $(a,b,c) = (a',b',c')$.

PROOF From the hypothesis it follows, since w depends only on the discriminant, that $w = w'$. Thus

$$\frac{-b+w}{a} = \omega = \omega' = \frac{-b'+w'}{a'} \qquad \text{so} \qquad \frac{-b+w}{a} = \frac{-b'+w}{a'}$$

and

$$\frac{-b-w}{a} = \eta = \eta' = \frac{-b'-w'}{a'} \qquad \text{so} \qquad \frac{-b-w}{a} = \frac{-b'-w}{a'} .$$

Subtracting corresponding parts of these second equalities, we have

$$\frac{2w}{a} = \frac{2w}{a'} .$$

*The symbol $\sqrt{a}$ denotes the positive square root of the positive real number a.

6.56.2

So a = a´. From (-b+w)/a = (-b´+w)/a´ and a = a´ we infer b = b´.
Finally, from $-ac = \Delta-b^2 = \Delta´-b´^2 = -a´c´ = -ac´$ it follows that
c = c´.

<u>COMMENT</u> If Δ is not the square of an integer, then either $\Delta < 0$ and w
has the form $w = i\theta$ with $\theta > 0$, or $\Delta > 0$ and w is a real irrational
number. In either case, the hypothesis $\{^{\Delta = \Delta´}_{\omega = \omega´}$ implies again w = w´
and then that a´(-b+w) = a(-b´+w) so ab´-a´b = (a-a´)w. Since w is
either complex or irrational, it follows that ab´-a´b and a-a´ both
vanish, which gives a = a´ and b = b´, from which it follows as
before that c = c´. Hence, if Δ is not the square of an integer then
$\{^{\Delta = \Delta´}_{\omega = \omega´}$ implies (a,b,c) = (a´,b´,c´).

<u>PROPOSITION 3</u> Let f = (a,b,c) ~ (a´,b´,c´) = f´ and aa´ $\neq$ 0 and
$f\begin{pmatrix}\alpha & \beta \\ \gamma & \delta\end{pmatrix} = f´.$ Then we will have
$$\omega = \frac{\alpha\omega´+\beta}{\gamma\omega´+\delta} \qquad \eta = \frac{\alpha\eta´+\beta}{\gamma\eta´+\delta}.$$

<u>Proof</u> From the hypothesis we have the identity in x´,y´:
$$a(\alpha x´+\beta y´)^2 + 2b(\alpha x´+\beta y´)(\gamma x´+\delta y´) + c(\gamma x´+\delta y´)^2$$
$$= a´x´^2 + 2b´x´y´ + c´y´^2. \tag{56.1}$$

We introduce x´ = $\omega´$, y´ = 1 into it and get the equation
$$a(\alpha\omega´+\beta)^2 + 2b(\alpha\omega´+\beta)(\gamma\omega´+\delta) + c(\gamma\omega´+\delta)^2$$
$$= a´\omega´^2 + 2b´\omega´ + c´ = 0. \tag{56.2}$$

I say that $\gamma\omega´+\delta \neq 0$. Indeed, if $\gamma\omega´+\delta = 0$ then by (56.2), we would
have $\alpha\omega´+\beta = 0$. But from
$$\alpha\omega´+\beta = 0$$
$$\gamma\omega´+\delta = 0$$

we would have $0 = \delta(\alpha\omega´+\beta)-\beta(\gamma\omega´+\delta) = (\alpha\delta-\beta\gamma)\omega´$ so $\omega´ = 0$ since $\alpha\delta-\beta\gamma = 1$.
Then we would have

6.56.3

$$\alpha \cdot 0 + \beta = 0$$
$$\gamma \cdot 0 + \delta = 0,$$

so $\beta = \delta = 0$ which contradicts $\alpha\delta - \beta\gamma = 1$.

Hence we may divide (56.2) by $(\gamma\omega'+\delta)^2$, and get the equation

$$a\left(\frac{\alpha\omega'+\beta}{\gamma\omega'+\delta}\right) + 2b\left(\frac{\alpha\omega'+\beta}{\gamma\omega'+\delta}\right) + c = 0,$$

which tells us that $\frac{\alpha\omega'+\beta}{\gamma\omega'+\delta}$ is a root of $f = (a,b,c)$. In the same way, we find that $\frac{\alpha\eta'+\beta}{\gamma\eta'+\delta}$ is a root of $f = (a,b,c)$. Thus it only remains to show that $\frac{\alpha\omega' + \beta}{\gamma\omega' + \delta}$ is the <u>first</u> root ω:

$$\omega = \frac{\alpha\omega'+\beta}{\gamma\omega'+\delta}, \qquad \text{that is } \omega = S\omega' \text{ with* } S = \begin{pmatrix} \alpha & \beta \\ \gamma & \delta \end{pmatrix}.$$

Let us assume that this is not true; that

$$\frac{\alpha\omega'+\beta}{\gamma\omega'+\delta} = \eta \qquad \text{so} \qquad \frac{\alpha\eta'+\beta}{\gamma\eta'+\delta} = \omega.$$

We form the difference:

$$\eta - \omega = \frac{\alpha\omega'+\beta}{\gamma\omega'+\delta} - \frac{\alpha\eta'+\beta}{\gamma\eta'+\delta} =$$
$$= \frac{(\alpha\omega'+\beta)(\gamma\eta'+\delta) - (\alpha\eta'+\beta)(\gamma\omega'+\delta)}{(\gamma\omega'+\delta)(\gamma\eta'+\delta)}. \tag{56.3}$$

But

$$(\alpha\omega'+\beta)(\gamma\eta'+\delta) - (\alpha\eta'+\beta)(\gamma\omega'+\delta) = \alpha\delta(\omega'-\eta') - \beta\gamma(\omega'-\eta')$$
$$= (\alpha\delta - \beta\gamma)(\omega'-\eta') = \omega'-\eta'$$

so

$$\eta - \omega = \frac{\omega' - \eta'}{(\gamma\omega'+\delta)(\gamma\eta'+\delta)}.$$

Moreover, we have

$$\eta - \omega = \frac{-2w}{a} \qquad \omega' - \eta' = \frac{2w'}{a'} = \frac{2w}{a'}$$

(since $f \sim f'$ implies $\Delta = \Delta'$ and thus $w = w'$).

Hence

$$\frac{-2w}{a} = \eta - \omega = \frac{\omega' - \eta'}{(\gamma\omega'+\delta)(\gamma\eta'+\delta)} = \frac{2w}{a'(\gamma\omega'+\delta)(\gamma\eta'+\delta)},$$

*It will prove convenient to also symbolize the Möbius transformation $\omega = \frac{\alpha\omega'+\beta}{\gamma\omega'+\delta}$ with S, writing $\omega = S\omega'$.

6.56.4

and then, after division by $2w$

$$\frac{-1}{a} = \frac{1}{a'(\gamma\omega'+\delta)(\gamma\eta'+\delta)} \ . \tag{56.4}$$

We recall now the identity in x',y':

$$a'x'^2+2b'x'y'+c'y'^2 = a'(x'-\omega'y')(x'-\eta'y'),$$

and thus, according to (56.1),

$$a(\alpha x'+\beta y')^2+2b(\alpha x'+\beta y')(\gamma x'+\delta y') +c(\gamma x'+\delta y')^2 = a'(x'-\omega'y')(x'-\eta'y').$$

Into this we introduce the values $x' = \delta$, $y' = -\gamma$ and get the equation

$$a(\alpha\delta-\beta\gamma)^2+2b(\alpha\delta-\beta\gamma)(\gamma\delta-\delta\gamma)+c(\gamma\delta-\delta\gamma)^2 = a'(\delta+\omega'\gamma)(\delta+\eta'\gamma)$$

and so

$$a = a'(\gamma\omega'+\delta)(\gamma\eta'+\delta),$$

which, combined with (56.4) gives

$$\frac{-1}{a} = \frac{1}{a} \ .$$

This is impossible, since $1/a \neq 0$. Consequently it must be that

$$\omega = \frac{\alpha\omega'+\beta}{\gamma\omega'+\delta} \qquad \text{and} \qquad \eta = \frac{\alpha\eta'+\beta}{\gamma\eta'+\delta}$$

as required.

<u>PROPOSITION 4</u>. If the discriminant of $f = (a,b,c)$ is equal to the discriminant of $f' = (a',b',c')$ and if $aa' \neq 0$ and $\omega = S\omega'$, $\eta = S\eta'$ where S is a unimodular substitution, then $fS = f'$.

<u>Proof</u> Let $fS = \bar{f} = (\bar{a},\bar{b},\bar{c})$ and $\omega = S\omega' = \frac{\alpha\omega'+\beta}{\gamma\omega'+\delta}$, $\eta = S\eta' = \frac{\alpha\eta'+\beta}{\gamma\eta'+\delta}$.

If $\gamma \neq 0$ then

$$\omega = \frac{1}{\gamma}\frac{\alpha\gamma\omega'+\alpha\delta-\alpha\delta+\beta\gamma}{\gamma\omega'+\delta} = \frac{\alpha}{\gamma} - \frac{\alpha\delta-\beta\gamma}{\gamma(\gamma\omega'+\delta)} \ ,$$

so $\omega \neq \frac{\alpha}{\gamma}$ and similarly for η. Then

$$\bar{a} = f(\alpha,\gamma) = a\alpha^2+2b\alpha\gamma+c\gamma^2 \neq 0$$

since

6.56.5

$$\frac{\overline{a}}{\gamma^2} = a\left(\frac{\alpha}{\gamma}\right)^2 + 2b\frac{\alpha}{\gamma} + c \neq 0$$

because $\omega \neq \frac{\alpha}{\gamma}$ and $\eta \neq \frac{\alpha}{\gamma}$ and ω, η are the roots of $ax^2+2bx+c = 0$. If $\gamma = 0$ then $\overline{a} = f(\alpha,\gamma) = a\alpha^2 \neq 0$, since $a \neq 0$ and $\alpha \neq 0$ because $\alpha = \gamma = 0$ would result in $1 = \alpha\delta-\beta\gamma = 0$.

Thus $\overline{f} = fS$ is quadratic form with first coefficient $\overline{a} \neq 0$. According to Proposition 3, $\omega = S\overline{\omega}$, $\eta = S\overline{\eta}$. Since, by hypothesis, $\omega = S\omega´$ and $\eta = S\eta´$, we have $\overline{\omega} = \omega´$ and $\overline{\eta} = \eta´$. Moreover discriminant $(\overline{f}) =$ discriminant $(f) =$ discriminant $(f´)$. Consequently, according to Proposition 2, $\overline{f} = f´$, so $f´ = \overline{f} = fS$ as required.

57. UNIMODULAR EQUIVALENCE OF A QUADRATIC FORM WITH ITSELF; THE EQUATION OF
FERMAT (AND OF PELL AND LAGRANGE)

Let f be a positive form. If it is not reduced, then, using the procedure outlined in §54, we may determine an equivalent reduced form ϕ. Let S_0 be the unimodular substitution emerging from this procedure which transforms f into ϕ: $fS_0 = \phi$. If f is reduced, then it is equivalent to itself and for S_0 we may take $\left(\begin{smallmatrix} 1 & 0 \\ 0 & 1 \end{smallmatrix}\right)$ or $\left(\begin{smallmatrix} -1 & 0 \\ 0 & -1 \end{smallmatrix}\right)$.

We pose the problem: What unimodular substitutions S other than S_0 transform f into ϕ?*

From $fS_0 = \phi$ and $fS = \phi$ we conclude that $fS_0 = fS$ and then, since $f = fSS_0^{-1}$, that SS_0^{-1} transforms f into itself. If we know all the unimodular substitutions T for which $fT = f$, then, determining from $SS_0^{-1} = T$ that $S = TS_0$, we will evidently find all the unimodular substitutions S which transform f into ϕ. Indeed $f(TS_0) = (fT)S_0 = fS_0 = \phi$. Thus we end up with the problem:

Determine all the unimodular substitutions T for which $fT = f$.

We now treat this problem. Let $f = (a,b,c)$ be an arbitrary quadratic form with first coefficient $a \neq 0$ and let $f\left(\begin{smallmatrix} \alpha & \beta \\ \gamma & \delta \end{smallmatrix}\right) = f$ with $\alpha\delta - \beta\gamma = 1$. In accordance with propositions 3 and 4 of the previous section, a necessary and sufficient condition for this is that

$$\omega = \frac{\alpha\omega + \beta}{\gamma\omega + \delta} \qquad \eta = \frac{\alpha\eta + \beta}{\gamma\eta + \delta} \, .$$

*The reader will recall from §53 the importance of this problem for the
solution of the second problem of §46 for binary quadratic forms.

6.57.2

Thus, it is necessary and sufficient that

$$\gamma\omega^2 + (\delta-\alpha)\omega - \beta = 0 \quad \text{and} \quad \gamma\eta^2 + (\delta-\alpha)\eta - \beta = 0.$$

The equation $\gamma x^2 + (\delta-\alpha)x - \beta = 0$ will then have the same roots ω, η as the equation $ax^2 + 2bx + c = 0$, and so the same roots as

$$\frac{a}{\sigma} x^2 + \frac{2b}{\sigma} x + \frac{c}{\sigma} = 0$$

where σ is the greatest common divisor of a, 2b, c. Thus it is necessary and sufficient to have

$$\gamma = u \frac{a}{\sigma} \qquad \delta-\alpha = u \frac{2b}{\sigma} \qquad -\beta = u \frac{c}{\sigma} \tag{57.1}$$

where u is a rational number since γ and a/σ are integers. Indeed, u will be an integer, for let $u = k/\ell$, $\ell > 0$, be in lowest terms. Then we will have

$$\gamma = \frac{k}{\ell} \frac{a}{\sigma} \qquad \delta-\alpha = \frac{k}{\ell} \frac{2b}{\sigma} \qquad \beta = -\frac{k}{\ell} \frac{c}{\sigma}$$

so

$$\ell\gamma = k \frac{a}{\sigma} \qquad \ell(\delta-\alpha) = k \frac{2b}{\sigma} \qquad \ell\beta = -k \frac{c}{\sigma}.$$

Since $(\ell,k) = 1$, ℓ is a divisor of a/σ, $2b/\sigma$, c/σ and thus ℓ is a divisor of the greatest common divisor of a/σ, $2b/\sigma$, c/σ which is 1; so $\ell = 1$ and $u = k = $ integer.

Now we set

$$\delta+\alpha = \frac{2t}{\sigma} \tag{57.2}$$

where t is some rational number. From (57.1) and (57.2) we infer

$$2\delta = \frac{2t}{\sigma} + \frac{2bu}{\sigma} \qquad \text{so} \qquad \delta = \frac{t+bu}{\sigma}$$

and

$$2\alpha = \frac{2t}{\sigma} - \frac{2bu}{\sigma} \qquad \text{so} \qquad \alpha = \frac{t-bu}{\sigma}.$$

Thus we have found that

$$\alpha = \frac{t-bu}{\sigma} \qquad \beta = \frac{-cu}{\sigma}$$

$$\gamma = \frac{au}{\sigma} \qquad \delta = \frac{t+bu}{\sigma} .$$

However, $\alpha\delta - \beta\gamma = 1$ so

$$\frac{t^2 - b^2 u^2}{\sigma^2} + \frac{acu^2}{\sigma^2} = 1.$$

Hence

$$t^2 - \Delta u^2 = \sigma^2 .$$

From this last equation we conclude that t^2 is an integer, and since t is a rational number it must be an integer.

We have now reached the following result:

If $f\begin{pmatrix} \alpha & \beta \\ \gamma & \delta \end{pmatrix} = f$ then there exist two integers u and t which satisfy the relations

$$t^2 - \Delta u^2 = \sigma^2, \tag{57.3}$$

$$\alpha = \frac{t-bu}{\sigma} \qquad \beta = \frac{-cu}{\sigma}$$

$$\gamma = \frac{au}{\sigma} \qquad \delta = \frac{t+bu}{\sigma} , \tag{57.4}$$

where σ is the greatest common divisor of a, $2b$, c.

Conversely, let t and u be two integers which satisfy the equation (57.3) for which $\Delta = b^2 - ac$ with a,b,c integers, $a \neq 0$, and σ the greatest common divisor of a, $2b$, c. Then I assert that $\alpha, \beta, \gamma, \delta$ defined by formulas (57.4) are integers for which $\alpha\delta - \beta\gamma = 1$ and $f\begin{pmatrix} \alpha & \beta \\ \gamma & \delta \end{pmatrix} = f$.

6.57.4

We first prove that $\alpha,\beta,\gamma,\delta$ defined by (57.4) are integers. This is obvious for γ and β because a/σ and c/σ and u are integers. Moreover, we have

$$\delta-\alpha = \frac{2bu}{\sigma} = \text{integer},$$

and

$$\alpha\delta = \frac{t^2-b^2u^2}{\sigma^2} = \frac{\sigma^2+\Delta u^2-b^2u^2}{\sigma^2} = 1 - \frac{acu^2}{\sigma^2} = 1 + \gamma\beta = \text{integer}.$$

Hence

$$(\delta+\alpha)^2 = (\delta-\alpha)^2 + 4\alpha\delta = \text{integer},$$

so $\delta+\alpha$ is an integer. Also $(\delta+\alpha)^2 \equiv (\delta-\alpha)^2 \pmod 4$, so, setting

$$\delta-\alpha = g \tag{57.5}$$

$$\delta+\alpha = g',$$

we have g and g' are integers and $g^2 \equiv g'^2 \pmod 4$.

This congruence may be written $(g-g')(g+g') \equiv 0 \pmod 4$, and tells us that the two integers $g-g'$ and $g+g'$ cannot both be odd. If $g-g' \equiv 0 \pmod 2$ then $g+g' \equiv g-g'+2g' \equiv 0 \pmod 2$ and if $g+g' \equiv 0 \pmod 2$ then $g-g' \equiv g+g'-2gg' \equiv 0 \pmod 2$.

From this and (57.5) it now follows that

$$\delta = \frac{g+g'}{2} = \text{integer} \quad \text{and} \quad \alpha = \frac{g'-g}{2} = \text{integer}.$$

Next we note that

$$\alpha\delta-\beta\gamma = \frac{t^2-b^2u^2}{\sigma^2} + \frac{acu^2}{\sigma^2} = \frac{t^2-\Delta u^2}{\sigma^2} = \frac{\sigma^2}{\sigma^2} = 1.$$

It remains for us to prove that

$$f\left(\begin{matrix} \dfrac{t-bu}{\sigma} & -\dfrac{cu}{\sigma} \\ \dfrac{au}{\sigma} & \dfrac{t+bu}{\sigma} \end{matrix}\right) = f. \qquad (57.6)$$

Our first proof of this is from the following three calculations

$$f\left(\frac{t-bu}{\sigma}, \frac{au}{\sigma}\right) = \frac{1}{\sigma^2}\left[a(t-bu)^2 + 2b(t-bu)au + ca^2u^2\right]$$

$$= \frac{a}{\sigma^2}\left[t^2 - \Delta u^2\right] = \frac{a}{\sigma^2}\,\sigma^2 = a$$

$$f\left(\frac{t-bu}{\sigma}, \frac{au}{\sigma}\middle|\frac{-cu}{\sigma}, \frac{t+bu}{\sigma}\right) = \frac{1}{\sigma^2}\left[-a(t-bu)cu - b(acu^2 - t^2 + b^2u^2) + cau(t+bu)\right]$$

$$= \frac{b}{\sigma^2}\left[acu^2 - acu^2 + t^2 - b^2u^2 + acu^2\right] = \frac{b}{\sigma^2}\,\sigma^2 = b$$

$$f\left(\frac{-cu}{\sigma}, \frac{t+bu}{\sigma}\right) = \frac{1}{\sigma^2}\left[ac^2u^2 - 2bcu(t+bu) + c(t+bu)^2\right]$$

$$= \frac{c}{\sigma^2}\left[t^2 - \Delta u^2\right] = \frac{c}{\sigma^2}\,\sigma^2 = c.$$

Our second proof of (57.6) is furnished by the observation that the steps from the hypothesis $f\left(\begin{smallmatrix}\alpha & \beta\\ \gamma & \delta\end{smallmatrix}\right) = f$ to (57.3) and (57.4) are each reversible. Finally a third proof by an entirely different method is as follows:

We have $\delta - \alpha = \dfrac{2bu}{\sigma}$ and $\gamma = \dfrac{au}{\sigma}$, $\beta = \dfrac{-cu}{\sigma}$. So we have the identity with respect to x

$$\gamma x^2 + (\delta-\alpha)x - \beta = \frac{u}{\sigma}(ax^2 + 2bx + c) = \frac{u}{\sigma}(x-\omega)(x-\eta).$$

Hence

$$\omega(\gamma\omega+\delta) - (\alpha\omega+\beta) = \gamma\omega^2 + (\delta-\alpha)\omega - \beta = 0$$

$$\eta(\gamma\eta+\delta) - (\alpha\eta+\beta) = \gamma\eta^2 + (\delta-\alpha)\eta - \beta = 0,$$

so

$$\omega = \frac{\alpha\omega+\beta}{\gamma\omega+\delta} \quad\text{and}\quad \eta = \frac{\alpha\eta+\beta}{\gamma\eta+\delta}.$$

6.57.6

Consequently, by proposition 4 of the preceeding section, we have

$$f\begin{pmatrix} \alpha & \beta \\ \gamma & \delta \end{pmatrix} = f.$$

Let us note that the equation $t^2 - \Delta u^2 = \sigma^2$ has in bibliographies

one of the three names: Equation of Fermat or of Pell or of Lagrange.

6.58.1

58. THE DIVISORS OF A QUADRATIC FORM

Let $f = ax^2+2bxy+cy^2 = (a,b,c)$. We call the greatest common divisor σ of a, 2b, c the divisor of f. The greatest common divisor τ of a,b,c we call the paradivisor of f.*

Since $\tau|\sigma$, $\tau \leqslant \sigma$.

<u>PROPOSITION 1</u> Each integer representable by f is divisible by the divisor σ of the form f.

The proof is immediate.

<u>PROPOSITION 2</u> Two equivalent quadratic forms (a,b,c) and (a',b',c') have the same divisor σ and the same paradivisor τ.

<u>Proof</u> Let $(a,b,c)\begin{pmatrix} \alpha & \beta \\ \gamma & \delta \end{pmatrix} = (a',b',c')$. Then, as we know,
$a' = a\alpha^2+2b\alpha\gamma+c\gamma^2$, $b' = a\alpha\beta+b(\alpha\delta+\beta\gamma)+c\gamma\delta$, $c' = a\beta^2+2b\beta\delta+c\delta^2$.
Consequently σ is a divisor of $a',2b',c'$ and τ is a divisor of a',b',c'. Thus $\sigma|\sigma'$ and $\tau|\tau'$ where σ' and τ' are, respectively, the divisor and paradivisor of (a',b',c').

Conversely $(a',b',c')S^{-1} = (a,b,c)$ where $S = \begin{pmatrix} \alpha & \beta \\ \gamma & \delta \end{pmatrix}$. According to what we proved above, $\sigma'|\sigma$ and $\tau'|\tau$. From $\sigma|\sigma'$ and $\sigma'|\sigma$ it follows that $\sigma = \sigma'$ and similarly $\tau = \tau'$.

<u>PROPOSITION 3</u> If $2b/\sigma$ is even, then $\sigma = \tau$. If $2b/\sigma$ is odd then $\sigma = 2\tau$.

<u>Proof</u> The integers a/σ, $2b/\sigma$, c/σ have greatest common divisor 1.
If $2b/\sigma$ is even, then b/σ is an integer, so σ is a divisor of b, and thus of τ. Since $\tau|\sigma$, we have $\sigma = \tau$.

---

*There seems to be no commonly accepted English term for τ. I have adapted a Greek term.

6.58.2

If $2b/\sigma$ is odd then $2b/\sigma = 2k+1$ and $2b = (2k+1)\sigma$, so σ is even and $\sigma/2$ an integer. Then the greatest common divisor of a/σ, $b/(\sigma/2)$, c/σ will be 1 and $b/(\sigma/2)$ will be odd. So the greatest common divisor of $2a/\sigma$, $b/(\sigma/2)$, $2c/\sigma$, that is, of $a/(\sigma/2)$, $b/(\sigma/2)$, $c/(\sigma/2)$ will be 1, and consequently $\tau = \sigma/2$ which implies $\sigma = 2\tau$.

DEFINITION If $\tau = 1$ then $f = (a,b,c)$ is called a _primitive_ quadratic form.

We note that in our investigation of which integers are represented by quadratic forms we may limit ourselves to primitive quadratic forms.

In fact, if $ax^2+2bxy+cy^2 = m$ with paradivisor $\tau > 1$ then, setting $a = a_1\tau$, $b = b_1\tau$, $c = c_1\tau$ we get

$$\tau(a_1x^2+2b_1xy+c_1y^2) = m,$$

so

$$a_1x^2+2b_1xy+c_1y^2 = m/\tau.$$

If we find the integers which are representable by the primitive quadratic form $a_1x^2+2b_1xy+c_1y^2$, then, multiplying these integers by τ we obtain the integers representable by $ax^2+2bxy+cy^2$.

DEFINITION Let $f = (a,b,c)$ be primitive ($\tau = 1$). Then we have $\sigma = \tau$ or $\sigma = 2\tau$. In the case $\sigma = \tau = 1$, f is called _primitive of the first kind_, and in the case $\sigma = 2\tau = 2$, f is called _primitive of the second kind_*

For example, the quadratic form $(3,2,5)$ is primitive of the first kind and $(2,1,10)$ is primitive of the second kind.

---

*Terminology on this matter is not fixed and the reader should check the definitions for each work he consults.

6.58.3

Indeed, like (2,1,10), a form will be primitive of the second kind if and only if a and c are even and b is odd. In this case we will have $\Delta = b^2 - ac \equiv 1 \pmod 4$.

Moreover, if $\Delta \equiv 1 \pmod 4$ then the form $2x^2 + 2xy + \frac{1-\Delta}{2}\, y^2$ is primitive of the second kind, and has discriminant Δ. Thus there exist primitive forms of the second kind having a given discriminant Δ if and only if $\Delta \equiv 1 \pmod 4$.

<u>REMARK</u> If $(a,b,c)\begin{pmatrix}\alpha & \beta \\ \gamma & \delta\end{pmatrix} = (a´,b´,c´)$ where $\begin{pmatrix}\alpha & \beta \\ \gamma & \delta\end{pmatrix}$ is a unimodular substitution then, as we have seen, the divisors and paradivisors of the two forms are the same: $\tau = \tau´$. Thus we also have

$$\left(\frac{a}{\tau}, \frac{b}{\tau}, \frac{c}{\tau}\right) \begin{pmatrix}\alpha & \beta \\ \gamma & \delta\end{pmatrix} = \left(\frac{a´}{\tau}, \frac{b´}{\tau}, \frac{c´}{\tau}\right).$$

In other words, from the equivalence $(a,b,c) \sim (a´,b´,c´)$ it follows that $(a/\tau,\ b/\tau,\ c/\tau) \sim (a´/\tau,\ b´/\tau,\ c´/\tau)$ which constitutes an equivalence of primitive quadratic forms. Thus to find all transformations S so that $fS = f$ it suffices to consider the primitive f.

59. EQUIVALENCE OF A FORM WITH ITSELF AND SOLUTION OF THE EQUATION OF FERMAT FOR FORMS WITH NEGATIVE DISCRIMINANT Δ

Let (a,b,c) be primitive with discriminant $\Delta = -D < 0$. We have to solve the equation $t^2 + Du^2 = \sigma^2$.

1st CASE $\sigma = 1$. Let first $D = 1$, so the equation becomes $t^2 + u^2 = 1$. The solutions are obviously $(u = 0, t = \pm 1)$ or $(u = \pm 1, t = 0)$.

With $u = 0$ we have the substitutions

$$\alpha = \frac{t-bu}{\sigma} = \pm 1 \qquad \beta = 0$$

$$\gamma = 0 \qquad \delta = \frac{t+bu}{\sigma} = \pm 1,$$

whence $\begin{pmatrix} \alpha & \beta \\ \gamma & \delta \end{pmatrix}$ is either $\begin{pmatrix} 1 & 0 \\ 0 & 1 \end{pmatrix}$ or $\begin{pmatrix} -1 & 0 \\ 0 & -1 \end{pmatrix}$.

With $t = 0$, $u = \pm 1$ we have

$$\alpha = \mp b \qquad \beta = \mp c$$

$$\gamma = \pm a \qquad \delta = \pm b$$

and $\begin{pmatrix} \alpha & \beta \\ \gamma & \delta \end{pmatrix}$ is $\begin{pmatrix} -b & -c \\ a & b \end{pmatrix}$ or $\begin{pmatrix} b & c \\ -a & -b \end{pmatrix}$.

In total, we have $2+2 = 4$ substitutions, of which two are

$\begin{pmatrix} 1 & 0 \\ 0 & 1 \end{pmatrix}$ and $\begin{pmatrix} -1 & 0 \\ 0 & -1 \end{pmatrix}$.

Secondly, let $D > 1$. Then $u = 0$ and $t = \pm 1$, and we get only the two substitutions $\begin{pmatrix} 1 & 0 \\ 0 & 1 \end{pmatrix}$ and $\begin{pmatrix} -1 & 0 \\ 0 & -1 \end{pmatrix}$.

2nd CASE $\sigma = 2$. Then the quadratic form (a,b,c) is primitive of the second kind, so $\Delta \equiv 1 \pmod 4$ and thus $D = -\Delta \equiv -1 \equiv 3 \pmod 4$. The possible values of D are the following

$$D = 3, 7, 11, \ldots$$

If $D > 3$, that is $D \geqslant 7$, then the equation $t^2+Du^2 = \sigma^2 = 4$ has only the solution $(u = 0, t = \pm 2)$. The corresponding substitutions are

$$\alpha = \frac{\pm 2}{2} \quad \beta = 0$$
$$\gamma = 0 \quad \delta = \frac{\pm 2}{2}$$

that is $\begin{pmatrix} \pm 1 & 0 \\ 0 & \pm 1 \end{pmatrix}$.

If $D = 3$, then the equation of Fermat becomes $t^2+3u^2 = 4$. Thus the solutions must have $u = 0$ or $u = \pm 1$. With $u = 0$ we have $t = \pm 2$ and the corresponding substitutions are those we found above: $\begin{pmatrix} \pm 1 & 0 \\ 0 & \pm 1 \end{pmatrix}$. With $u = \pm 1$ we have $t = \pm 1$ with 4 possible arrangements of sign. So the corresponding substitutions are

$$\begin{pmatrix} \dfrac{1-b}{2} & -\dfrac{c}{2} \\ \dfrac{a}{2} & \dfrac{1+b}{2} \end{pmatrix} \quad \begin{pmatrix} \dfrac{-1+b}{2} & \dfrac{c}{2} \\ -\dfrac{a}{2} & \dfrac{-1-b}{2} \end{pmatrix}$$

$$\begin{pmatrix} \dfrac{-1-b}{2} & -\dfrac{c}{2} \\ \dfrac{a}{2} & \dfrac{-1+b}{2} \end{pmatrix} \quad \begin{pmatrix} \dfrac{1+b}{2} & \dfrac{c}{2} \\ -\dfrac{a}{2} & \dfrac{1-b}{2} \end{pmatrix}.$$

In total, for $\Delta = -3$ the equation of Fermat has $2+4 = 6$ solutions. Thus we have

PROPOSITION Let k be the number of unimodular substitutions which transform a primitive quadratic form with negative discriminant into itself. Then $k = 2$ except

if $\Delta = -1$, $D = 1$, $\sigma = 1$ then $k = 4$,

if $\Delta = -3$, $D = 3$, $\sigma = 2$ then $k = 6$.

60. THE PRIMITIVE REPRESENTATIONS OF AN ODD INTEGER m BY x^2+y^2.

The quadratic form $f(x,y) = x^2+y^2$ is positive and has discriminant -1. We ask: if m is a positive integer, how many primitive representations by f has it? Since we are seeking primitive representations (x,y) we will have $(x,y) = 1$, so x and y cannot both be even. When m is odd one of x,y will be even and the other odd. When x and y are both odd then $m \equiv 2 \pmod 4$; in fact $m \equiv 2 \pmod 8$. Below we will concern ourselves with the case of <u>odd</u> m > 1.

To find the primitive representations of m by $f(x,y) = x^2+y^2$, should they exist, we must, according to the preceding sections (see §53), solve the congruence $z^2 \equiv \Delta = -1 \pmod m$. Let $m = p_1^{\alpha_1} p_2^{\alpha_2} \ldots p_k^{\alpha_k}$ where the p_j are k distinct primes > 2. The congruence $z^2 \equiv \Delta = -1 \pmod m$ is solvable if and only if the congruence $z^2 \equiv -1 \pmod{p_j}$ is solvable for each p_j $(j = 1,\ldots,k)$. Thus it is necessary and sufficient that

$$\left(\frac{-1}{p_1}\right) = \left(\frac{-1}{p_2}\right) = \ldots = \left(\frac{-1}{p_k}\right) = 1,$$

so

$$p_j \equiv 1 \pmod 4 \text{ for } j = 1,\ldots,k.$$

Then there exist $\lambda = 2^k$ distinct solutions of $z^2 \equiv -1 \pmod m$ in any complete system of residues mod m. Let

$$n_1, n_2, \ldots, n_\lambda$$

be these solutions. We form the quadratic forms

$$\phi_j = \left(m, n_j, \frac{n_j^2+1}{m}\right) \text{ with discriminant } -1, (j = 1,\ldots,\lambda).$$

6.60.2

As we know, there exists only one reduced form with discriminant
$\Delta = -D = -1$. Consequently we have

$$\phi_i \sim \phi_j \text{ for } 1 < i, j < \lambda \text{ and } \phi_j \sim f, (j = 1,\ldots,\lambda).$$

According to the Proposition of the preceding section, there exist four substitutions S for which $\phi_j S = f$, $j = 1,\ldots,\lambda$. Hence there exist in total $4 \cdot 2^k = 2^{k+2}$ primitive representations of m. However, we note the following:

If (α,γ) is a primitive representation of m, $m = \alpha^2+\gamma^2$, then along with it come eight related representations

$$(x = \pm\alpha, y = \pm\gamma) \text{ and } (x = \pm\gamma, y = \pm\alpha),$$

which are distinct since $\alpha \neq \gamma$ because m is odd. These eight do not differ essentially from (α,γ). Consequently there are $2^{k+2}/2^3 = 2^{k-1}$ essentially different primitive representations of m.

Specializing to the case $k = 1$ and $m = p$, an odd prime, there exists essentially only one representation. In other words, a prime $p \equiv 1 \pmod 4$ may be decomposed into the sum of two squares in essentially only one way.

EXAMPLES The integer 3 cannot be represented by x^2+y^2; 5 may be represented in essentially one way $5 = 1^2+2^2$; 7 and 11 are congruent to 3 mod 4 and hence cannot be represented by x^2+y^2; $13 \equiv 1 \pmod 4$ and can be represented in essentially only one way: $13 = 2^2+3^2$; $65 = 5 \cdot 13$ has two essentially different representations, $65 = 4^2+7^2 = 1^2+8^2$.

61. THE REPRESENTATION OF AN INTEGER m BY A COMPLETE SYSTEM OF FORMS WITH GIVEN DISCRIMINANT $\Delta = -D < 0$

As we know, to a given discriminant $\Delta < 0$ there corresponds a finite system $f_1, f_2, \ldots, f_h$ of forms with the property that any form f with discriminant Δ is equivalent to some definite one of $f_1, f_2, \ldots, f_h$, which then represents the same numbers as f. We will refer to the system $f_1, \ldots, f_h$ of inequivalent forms as a complete system of forms for Δ.

Let m > 0 be an integer. We are interested in the representations of m by the system $f_1, \ldots, f_h$, that is, by the representations of m by any form of the system. Two representations (x', y') and (x'', y'') will count as different if the pairs (x', y') and (x'', y'') are distinct or if the pairs are identical but the representation is by different forms f_i and f_j $(i \neq j)$ of the system. With E(m) we designate the number of distinct primitive representations of m by the system, and by A(m) the number of representations of m, primitive or not, by the system.

We will assume that $(m, 2\Delta) = 1$, so m is odd and $(m, \Delta) = 1$. Then each solution of the congruence $z^2 \equiv \Delta \pmod{m}$ will satisfy $(z, \Delta) = 1$. We solve the congruence and determine all distinct solutions $n_1, \ldots, n_\lambda$ in a reduced system of residues mod m. Thus $(m, n_j) = 1$, as follows from $n_j^2 \equiv \Delta \pmod{m}$ and $(m, \Delta) = 1$. Then we form the quadratic forms

$$\phi_j = \left(m, \, n_j, \, \frac{n_j^2 - m}{m}\right) \qquad j = 1, \ldots, \lambda.$$

In them $(2n_j, m) = 1$ since m is odd. Thus we have $\sigma = \tau = 1$ for the ϕ_j $(j = 1, \ldots, \lambda)$ and the ϕ_j are primitive of the first kind. For each ϕ_j there exists one and only one $i \in \{1, \ldots, h\}$ such that

6.61.2

$$\phi_j \sim f_i.$$

Moreover there exist κ unimodular substitutions S such that $f_i S = f_i$. (The number κ is the same for each f_i since κ is the number solutions of the equation of Fermat: $t^2 + Du^2 = 1$.) Hence we will have

$$E(m) = \kappa\lambda.$$

Let now $m = p_1^{\alpha_1} p_2^{\alpha_2} \cdots p_k^{\alpha_k}$, where $(p_j, \Delta) = 1$. The congruence $z^2 \equiv \Delta \pmod{m}$ is solvable if and only if

$$\left(\frac{\Delta}{p_1}\right) = \left(\frac{\Delta}{p_2}\right) = \cdots = \left(\frac{\Delta}{p_k}\right) = 1.$$

If it is solvable the number of solutions (which all belong to a reduced system of residues) mod m is $\lambda = 2^k$, as we know from §36. If it is not solvable, the number of solutions is 0. In either case, solvable or not, we have

$$\lambda = \prod_{j=1}^{k} \left(1 + \left(\frac{\Delta}{p_j}\right)\right)$$

$$= \left(\frac{\Delta}{1}\right) + \sum_j \left(\frac{\Delta}{p_j}\right) + \sum_{i \neq j} \left(\frac{\Delta}{p_i p_j}\right) + \cdots + \left(\frac{\Delta}{p_1 p_2 \cdots p_k}\right)$$

(where $\left(\frac{\Delta}{1}\right)$ is always 1). So

$$\lambda = \sum_{\varepsilon \mid m} \left(\frac{\Delta}{\varepsilon}\right)$$

where ε runs through the divisors of m which themselves have no divisors which are squares $n^2 > 1$. Such divisors are called square-free divisors. Thus we have

$$E(m) = \kappa \sum_{\varepsilon \mid m} \left(\frac{\Delta}{\varepsilon}\right) = \begin{cases} \kappa 2^k \\ 0 \end{cases} \quad (\varepsilon \text{ square free}). \tag{61.1}$$

In general, $\kappa = 2$; only for $\Delta = -D = -1$ is $\kappa = 4$, because our formes are primitive of the first kind.

Let now $f = (a,b,c)$ be an f_i $(i = 1,\ldots,h)$ and
$$m = ax^2+2bxy+cy^2.$$

This time let $(x,y) = q$, so that the representation is not necessarily primitive. Then $(x,y) = (qx',qy')$ where $(x',y') = 1$. Thus we will have
$$ax'^2+2bx'y'+cy'^2 = \frac{m}{q^2} \; ;$$
that is, to every representation of m by (a,b,c) corresponds a primitive representation of m/q^2 by (a,b,c) where $(x,y) = q$. Thus
$$A(m) = \sum_{q^2|m} E(\tfrac{m}{q^2}). \tag{61.2}$$

However, according to (61.1),
$$E(\tfrac{m}{q^2}) = \kappa\!\!\sum_{\varepsilon|m/q^2} (\tfrac{\Delta}{\varepsilon}) = \kappa\!\!\sum_{\varepsilon q^2|m} (\tfrac{\Delta}{\varepsilon q^2}), \tag{61.3}$$

where ε runs through the square free divisors of m/q^2. From (61.2) and (61.3) it follows that
$$A(m) = \kappa\!\!\sum_{q^2|m}\;\sum_{\varepsilon q^2|m} (\tfrac{\Delta}{\varepsilon q^2}) = \kappa\!\!\sum_{\delta|m} (\tfrac{\Delta}{\delta}), \tag{61.4}$$

where δ runs through all the positive divisors of m, because each divisor of m may be written uniquely in the form εq^2 where ε is square free. In the case that m has no square divisors $q^2 > 1$, the square free divisors ε are identical with the divisors δ and thus
$$E(m) = A(m) \qquad \text{(m has no square free divisors)}. \tag{61.5}$$

__EXAMPLES__ I) $\Delta = -D = -1$. To this Δ, as we know (§55), corresponds a representative form $f = x^2+y^2$ with the property that any form with discriminant $\Delta = -1$ is equivalent to $f = x^2+y^2$. Now let $m > 1$ be odd (so that $(m,2) = 1$). According to (61.4) and the above note on the value of κ, we will have
$$A(m) = 4\sum_{\delta|m} (\tfrac{-1}{\delta}) = 4(M-N), \tag{61.6}$$

6.61.4

where M = (number of divisors of m which are congruent to 1 mod 4) and
N = (number of divisors of M which are congruent to 3 mod 4).

For example, $A(3) = 4(1-1) = 0$, $A(5) = 4(2-0) = 8$ (actually, all the representations of 5 are $(\pm 1, \pm 2)$ and $(\pm 2, \pm 1)$).

COMMENT Since $A(m) \geqslant 0$, from (61.6) it follows that for $m > 1$

$$M \geqslant N.$$

For example, the divisors $\delta \equiv 1 \pmod 4$ of 63 are 1, 9, 21 (M = 3) and the divisors $\delta \equiv 3 \pmod 4$ are 3, 7, 63 (N = 3). For 93 we have M = 2; (1,93) and N = 2; (3,31). The integers 63 and 93 cannot be represented by $x^2 + y^2$.

For 29, we have M = 2 and N = 0, so 29 is representable by $x^2 + y^2$. Indeed, we have the eight representations $(\pm 2, \pm 5)$ and $(\pm 5, \pm 2)$.

II) $\Delta = -D = -2$. Here also the number of classes of equivalent forms is 1 and for a system of representatives we may take $x^2 + 2y^2$. Let $m > 1$ and $(m,4) = 1$, so m is odd. According to (61.4) we have

$$A(m) = 2\sum_{\delta \mid m} \left(\frac{-2}{\delta}\right) = 2(M-N)$$

where

M = (number of divisors $\delta \equiv 1,3 \pmod 8$),

N = (number of divisors $\delta \equiv 5,7 \pmod 8$).

Specializing to the case when m = p, an odd prime, we have

$$A(p) = 2(M-N) = \begin{cases} 2(2-0) & \text{for } p \equiv 1,3 \pmod 8 \\ 2(1-1) & \text{for } p \equiv 5,7 \pmod 8. \end{cases}$$

Hence a prime p may be decomposed into a sum of the form $x^2 + 2y^2$ if and only if $p \equiv 1,3 \pmod 8$, and in this case there is essentially only one such decomposition.

6.61.5

Now let m = 9. Then M = 3 (1,3,9) and N = 0, so A(9) = 2 · (3-0) = 6. Indeed we have the 6 representations (±3,0) and (±1,±2).

III) Δ = -D = -3. We consider again integers m with $(m,2\Delta)$ = (m,6) = 1; that is integers which do not have 2 or 3 as factors. Here a complete system of representatives is $\{x^2+3y^2,\ 2x^2+2xy+2y^2\}$. But $2x^2+2xy+2y^2$ represents exclusively even numbers, and thus need not be considered here. From (61.4) we get

$$A(m) = 2\sum_{\delta|m} \left(\frac{-3}{\delta}\right) = 2\sum_{\delta|m} \left(\frac{-1}{\delta}\right)\left(\frac{3}{\delta}\right) = 2\sum_{\delta|m} \left(\frac{\delta}{3}\right) = 2(M-N),$$

where (see §44)

M = (number of $\delta \equiv 1 \mod 3$),

N = (number of $\delta \equiv 2 \mod 3$).

Specializing to the case m = p, an odd prime, we have δ = 1, p. Thus

if $p \equiv 1$ (mod 3) then A(p) = 2(2-0) = 4,

if $p \equiv 2$ (mod 3) then A(p) = 2(1-1) = 0.

Consequently, each prime of the form p = 3k+1 is representable, and essentially in only one way, by x^2+3y^2. For example, 7 = $(\pm 2)^2+3(\pm 1)^2$.

COMMENT Let m be odd and $m \not\equiv 0$ (mod 3). Then, since A(m) > 0, M = (number of divisors $\delta \equiv 1$ (mod 3)) > N = (number of divisors $\delta \equiv 2$ (mod 3)). For example, for m = 125 we have M = 2 (1,5) and N = 2 (25,125). So A(125) = 0 and 125 cannot be decomposed in the form x^2+3y^2. For m = 79, we have M = 2, N = 0 and A(79) = 2(2-0) = 4. Indeed 79 = $(\pm 2)^2+3(\pm 5)^2$.

We will now consider the quadratic forms which are primitive of the second kind: τ = 1, σ = 2. Then $a \equiv c \equiv 0$ (mod 2) and $b \equiv 1$ (mod 2).

Let

$$g_1, g_2, \ldots, g_h.$$

be a complete set of representatives of forms of the second kind with a given value of the discriminant $\Delta = -D < 0$. Since $a = 2a'$, $c = 2c'$, $b = 2\mu+1$, we have $\Delta \equiv (2\mu+1)^2 - 4a'c' \equiv 1 \pmod 4$. The forms under consideration represent only even integers $2m$.

Let us again designate by $E'(2m)$ and $A'(2m)$ the number of primitive representations and the total number of representations respectively of $2m$ by forms (a,b,c) of the second kind where $(m, 2\Delta) = 1$. This last condition requires that m be odd. We solve the congruence $z^2 \equiv \Delta \pmod{2m}$ and designate by $n_1, n_2, \ldots, n_\lambda$ the distinct solutions in a reduced system of residues mod $2m$. From $(m, 2\Delta) = 1$ and $n_j^2 \equiv \Delta \pmod{2m}$ it follows that $(2m, n_j) = 1$. We form the quadratic forms

$$\phi_j = \left(2m, \; n_j, \; \frac{n_j^2 - \Delta}{2m}\right) \qquad j = 1, \ldots, \lambda.$$

These quadratic forms are primitive of the second kind, for $(2m, n_j) = 1$, so $\tau = 1$ and since $n_j^2 \equiv 1 \pmod 4$ (because n_j is odd) and $\Delta \equiv 1 \pmod 4$ we have $n_j^2 - \Delta \equiv 0 \pmod 4$; hence $(n_j^2 - \Delta)/2m$ is even, so $\sigma = 2$.

There exists for each ϕ_j a unique $g_i \sim \phi_j$. Moreover, there exist κ unimodular substitutions transforming g_i into itself, each of which gives a distinct representation of $2m$. In general $\kappa = 2$, but for the case $\Delta = -3$ we have $\kappa = 6$.

Consequently

$$E'(2m) = \kappa\lambda. \tag{61.7}$$

Let $m = p_1^{\alpha_1} \ldots p_k^{\alpha_k}$ with p_i odd primes. The number of solutions of $x^2 \equiv \Delta \pmod{2m}$ is (see §36)

6.61.7

$$\lambda = [1+(\tfrac{\Delta}{p_1})][1+(\tfrac{\Delta}{p_2})]\ldots[1+(\tfrac{\Delta}{p_k})] = \begin{cases} 2^k \\ 0. \end{cases} \tag{61.8}$$

From (61.7) and (61.8) as before

$$E'(2m) = \kappa\!\!\sum_{\varepsilon\,|\,m} (\tfrac{\Delta}{\varepsilon}) \qquad (\varepsilon \text{ square free})$$

and then

$$A'(m) = \kappa\!\!\sum_{\delta\,|\,m} (\tfrac{\Delta}{\delta})$$

where ϕ runs through the positive divisors of m.

EXAMPLE $\Delta = -3$. Then $h' = 1$ and a complete set of representatives of

forms of the second kind is

$$2x^2+2xy+2y^2.$$

Now let m be an integer with $(m,6) = 1$. Then

$$A'(2m) = 6\sum_{\delta\,|\,m} (\tfrac{-3}{\delta}) = 6\sum_{\delta\,|\,m} (\tfrac{\delta}{3}) = 6(M-N),$$

where

$\quad$ M = (number of $\delta \equiv 1 \pmod 3$),

$\quad$ N = (number of $\delta \equiv 2 \pmod 3$).

If m = 7 then $\delta = 1 \equiv 1 \pmod 3$ and $\delta = 7 \equiv 1 \pmod 3$ so M = 2, N = 0 and

$A'(14) = 6\cdot 2 = 12$. We verify this:

$$14 = 2x^2+2xy+2y^2$$

$$\Longleftrightarrow \quad 28 = 4x^2+4xy+4y^2$$

$$= (2x+y)^2 + 3y^2.$$

From $3y^2 < 28$ follows $|y| < 3$ with y = 0 excluded. The possible values

of y are then $\pm 1, \pm 2, \pm 3$.

For $y = \pm 1$, $(2x\pm 1)^2 = 25 \Rightarrow 2x\pm 1 = \pm 5$.

For $y = +1$, $2x+1 = \pm 5 \Rightarrow x = 2, -3$ with corresponding representations

$\quad$ (2,1), (-3,1).

6.61.8

For $y = -1$, $2x-1 = \pm 5$ => $x = 3, -2$ with corresponding representations
$(3,-1)$, $(-2,-1)$.

For $y = \pm 2$, $(2x\pm 2)^2 = 28-12 = 16$ => $2x\pm 2 = \pm 4$.

For $y = +2$, $2x+2 = \pm 4$ => $x = 1,-3$ with corresponding representations
$(1,2)$, $(-3,2)$.

For $y = -2$, $2x-2 = \pm 4$ => $x = 3,-1$ with corresponding representations
$(3,-2)$, $(-1,-2)$.

For $y = \pm 3$, $(2x\pm 3)^2 = 28-27 = 1$ => $2x\pm 3 = \pm 1$.

For $y = +3$, $2x+3 = \pm 1$ => $x = -1, -2$ with corresponding representations
$(-1,3)$, $(-2,3)$.

For $y = -3$, $2x-3 = \pm 1$ => $x = 2,1$ with corresponding representations
$(2,-3)$, $(1,-3)$.

Altogether there are indeed twelve representations.

62. REGULAR CONTINUED FRACTIONS

A continued fraction

$$a_0 + \cfrac{b_1}{a_1 + \cfrac{b_2}{a_2 + \cfrac{\displaystyle\ddots\;+\;\cfrac{b_{n-1}}{a_{n-1} + \cfrac{b_n}{a_n}}}{}}}$$

will be called regular if

$b_1 = b_2 = \ldots = b_n = 1$; $a_0, a_1, \ldots, a_{n-1}$ are integers, and $a_1, a_2, \ldots, a_{n-1}$ are positive, and $a_n \geqslant 1$ is a real number.

In §24 we saw that any rational number $x_0 = a/m$ with $m \geqslant 1$ may be represented as a regular continued fraction.

Let now x_0 be any real number. We set

$$x_0 = q_1 + \frac{1}{x_1} \quad \text{where } q_1 = [x_0] \text{ and } 0 \leqslant \frac{1}{x_1} < 1 \text{ so } x_1 > 1$$

$$x_1 = q_2 + \frac{1}{x_2} \quad \text{where } q_2 = [x_1] \text{ and } 0 \leqslant \frac{1}{x_2} < 1 \text{ so } x_2 > 1$$

and, in general

$$x_{k-1} = q_k + \frac{1}{x_k} \quad \text{where } q_k = [x_{k-1}] \text{ and } 0 \leqslant \frac{1}{x_k} < 1 \text{ so } x_k > 1.$$

As long as $x_0, x_1, \ldots, x_{k-1}$ are not integers, the above process may be continued; however, if x_k is an integer then the process terminates. The above algorithm is completely determined when x_0 is given. In it, q_1 is an integer, $q_2, q_3, \ldots, q_k$ are positive integers, and $x_k > 1$ is a real number. With elimination of $x_1, x_2, \ldots, x_{k-1}$ in the above k equations, we find a representation x_0 as a regular continued fraction

$$x_0 = q_1 + \cfrac{1}{q_2 + \cfrac{1}{q_3 + \cfrac{\displaystyle\ddots\;+\;\cfrac{1}{q_k + \cfrac{1}{x_k}}}{}}}$$

which we will symbolize by $(q_1,q_2,\ldots,q_k,x_k)$. For example

$$x_0 = \frac{4}{3} = -2+\frac{2}{3} = -2+\frac{1}{3/2} = -2+\cfrac{1}{1+\frac{1}{2}} = (-2,1,2)$$

and, for $x_0 = \sqrt{2}$,

$$\sqrt{2} = 1+(\sqrt{2}-1) = 1+\cfrac{1}{1/(\sqrt{2}-1)} = 1+\cfrac{1}{\sqrt{2}+1} = 1+\cfrac{1}{2+(\sqrt{2}-1)}$$

$$= 1+\cfrac{1}{2+\cfrac{1}{2+(\sqrt{2}-1)}}$$

$$= 1+\cfrac{1}{2+\cfrac{1}{2+\cfrac{1}{2+(\sqrt{2}-1)}}} \qquad\qquad = 1+\cfrac{1}{2+\cfrac{1}{2+\cfrac{1}{2+\cfrac{1}{\sqrt{2}+1}}}}$$

and the process may be repeated without end. Thus we have

$\sqrt{2} = (1,2,2,\ldots,2,\sqrt{2}+1)$, where the three dots represent an arbitrary

sequence formed with the number 2.

As is obvious from the above, the algorithm which we have introduced

assigns to an irrational number x_0 a non-terminating sequence of regular

continued fractions

$$(q_1,q_2,\ldots,q_n,x_n) = x_0 \qquad n = 1,2,3,\ldots \ ,$$

where $x_n > 1$ is an irrational number. The $q_1,q_2,\ldots,q_n$ are called the

partial quotients, the x_n the _final denominator_ of the continued fraction

$(q_1,q_2,\ldots,q_n,x_n)$. Below we will prove that

$$\lim_{n\to\infty} (q_1,q_2,\ldots,q_n) = x_0.$$

Regarding the representation of rational numbers by means of regular

continued fractions, we add the following:

Let x_0 be a rational number and

$$x_0 = (q_1,q_2,\ldots,q_k).$$

If $k > 1$, then q_k is an integer and $q_k > 1$, so $q_k \geqslant 2$ and we may write it

as $q_k = (q_k-1)+(1/1)$ with $q_k-1 \geqslant 1$. Thus

6.62.3

$$x_0 = (q_1, q_2, \ldots, (q_k-1), 1).$$

If $k = 1$ then $x_0 = (x_0-1)+(1/1)$ and we may write

$$x_0 = (q_1) = ((q_1-1), 1).$$

Thus every rational number has two developments as a regular continued fraction.

Now let x_0 be irrational and

$$x_0 = (q_1, q_2, \ldots, q_k, x_k) \text{ for } k = 1, 2, 3, \ldots .$$

We set, as in §24, for $r \geqslant 3$

$$Z_1 = q_1, \ Z_2 = q_2 q_1 + 1, \ Z_3 = q_3 Z_2 + Z_1, \ldots, Z_r = q_r Z_{r-1} + Z_{r-2}, \ldots ,$$

$$N_1 = 1, \ N_2 = q_2 \quad , \ N_3 = q_3 N_2 + N_1, \ldots, N_r = q_r N_{r-1} + N_{r-2}, \ldots .$$

As we found in §24, for $r \geqslant 2$

$$(q_1, \ldots, q_r) = \frac{Z_r}{N_r} \quad \text{and} \quad Z_r N_{r-1} - N_r Z_{r-1} = (-1)^r.$$

From the last relation we have

$$(Z_r, N_r) = 1 \quad \text{for } r \geqslant 2.$$

This is also true for $r = 1$. Moreover, we have for the integer N_r $(r = 1, 2, 3, \ldots)$

$$0 < N_1 \leqslant N_2 < N_3 < N_4 < \ldots < N_r < N_{r+1} < \ldots, \tag{62.1}$$

so

$$N_r \geqslant r-1 \quad \text{for } r \geqslant 1 \text{ and } \lim_{r \to \infty} N_r = + \infty. \tag{62.2}$$

<u>PROPOSITION 1</u> For $k \geqslant 2$ we have

$$x_0 = (q_1, \ldots, q_k, x_k) = \frac{Z_k x_k + Z_{k-1}}{N_k x_k + N_{k-1}} \tag{62.3}$$

and

$$x_0 = \frac{Z_k}{N_k} + \frac{(-1)^{k+1}}{N_k(N_k x_k + N_{k-1})} . \tag{62.4}$$

<u>Proof</u>

$$x_0 = q_1 + \cfrac{1}{q_2 + \cfrac{1}{x_2}} = q_1 + \frac{x_2}{q_2 x_2 + 1}$$

6.62.4

$$= \frac{(q_1 q_2 + 1)x_2 + q_1}{q_2 x_2 + 1} = \frac{Z_2 x_2 + Z_1}{N_2 x_2 + N_1} .$$

The assertion (62.3) is thus true for $k = 2$. Now let us assume that for some $k-1 > 2$

$$x_0 = \frac{Z_{k-1} x_{k-1} + Z_{k-2}}{N_{k-1} x_{k-1} + N_{k-2}} . \qquad (62.5)$$

Then we will show that (62.3) is also valid. Indeed, by the hypothesis (62.5)

$$x_0 = (q_1, \ldots, q_k, x_k) = (q_1, \ldots, q_{k-1}, \frac{q_k x_k + 1}{x_k})$$

$$= \frac{Z_{k-1}[(q_k x_k + 1)/x_k] + Z_{k-2}}{N_{k-1}[(q_k x_k + 1)/x_k] + N_{k-2}}$$

$$= \frac{x_k Z_{k-1} q_k + Z_{k-1} + x_k Z_{k-2}}{x_k N_{k-1} q_k + N_{k-1} + x_k N_{k-2}}$$

$$= \frac{(q_k Z_{k-1} + Z_{k-2})x_k + Z_{k-1}}{(q_k N_{k-1} + N_{k-2})x_k + N_{k-1}}$$

$$= \frac{Z_k x_k + Z_{k-1}}{N_k x_k + N_{k-1}} .$$

Thus (62.3) has been proved by mathematical induction.

To prove (62.4) we note that according to (62.3) we have for $k > 2$

$$x_0 = \frac{Z_k}{N_k} + \left(\frac{Z_k x_k + Z_{k-1}}{N_k x_k + N_{k-1}} - \frac{Z_k}{N_k} \right)$$

$$= \frac{Z_k}{N_k} + \frac{N_k Z_k x_k + N_k Z_{k-1} - Z_k N_k x_k - Z_k N_{k-1}}{N_k(N_k x_k + N_{k-1})}$$

$$= \frac{Z_k}{N_k} - \frac{Z_k N_{k-1} - N_k Z_{k-1}}{N_k(N_k x_k + N_{k-1})}$$

$$= \frac{Z_k}{N_k} - \frac{(-1)^k}{N_k(N_k x_k + N_{k-1})}$$

$$= \frac{Z_k}{N_k} + \frac{(-1)^{k+1}}{N_k(N_k x_k + N_{k-1})} ,$$

which is what we wished to prove.

6.62.5

COROLLARY 1 For $k \geq 2$

$$|x_0 - \frac{Z_k}{N_k}| < \frac{1}{N_k^2} \, . \tag{62.6}$$

Proof We have

$$|x_0 - \frac{Z_k}{N_k}| = \frac{1}{N_k(N_k x_k + N_{k-1})} \, ,$$

where $x_k > 1$ and $N_{k-1} \geq 1$, so $N_k(N_k x_k + N_{k-1}) > N_k^2$.

COROLLARY 2 For irrational x_0 we have.

$$\lim_{k \to \infty} \frac{Z_k}{N_k} = \lim (q_1, \ldots, q_k) = x_0 \, .$$

  Proof This follows from (62.6) and (62.2).

PROPOSITION 2 Let $q_1, q_2, \ldots, q_k$ $(k \geq 2)$ be integers, q_1 arbitrary and $q_2, \ldots, q_k$ positive, and $x_k > 1$ a real number. The continued fraction

$$(q_1, q_2, \ldots, q_k, x_k)$$

has some definite value x. Then x has $(q_1, \ldots, q_k, x_k)$ for its regular continued fraction (of length k+1).

Proof Set

$$x_{k-1} = q_k + \frac{1}{x_k} \; ; \text{ since } x_k > 1, \; [x_{k-1}] = q_k \text{ and } x_{k-1} > 1,$$

$$x_{k-2} = q_{k-1} + \frac{1}{x_{k-1}} \; ; \text{ since } x_{k-1} > 1, \; [x_{k-2}] = q_{k-1}, \text{ and } x_{k-2} > 1,$$

$$x_{k-3} = q_{k-2} + \frac{1}{x_{k-2}} \; ; \text{ since } x_{k-2} > 1, \; [x_{k-3}] = q_{k-2} \text{ and } x_{k-3} > 1,$$

$$\cdots \cdots$$

$$x_1 = q_2 + \frac{1}{x_2} \; ; \text{ since } x_2 > 1, \; [x_1] = q_2 \text{ and } x_1 > 1$$

$$y = q_1 + \frac{1}{x_1} \; ; \text{ since } x_1 > 1, \; [y] = q_1 .$$

From the above relations with elimination of $x_1, \ldots, x_{k-1}$ it follows immediately that

$$y = q_1 + \cfrac{1}{q_2 + \cfrac{1}{q_3 + \cfrac{\cdots}{\cdots + q_k + \cfrac{1}{x_k}}}} = (q_1, \ldots, q_k, x_k) = x \, ,$$

and thus that $(q_1,q_2,\ldots,q_k,x_k)$ is the regular continued fraction of $y = x$.

COROLLARY 3 Let $x = (q_1,q_2,\ldots,q_k,x_k) = (q_1,\ldots,q_k,q_{k+1},\ldots,q_{k+\ell},x_\ell)$ with $k,\ell \geqslant 1$. Then $x_k = (q_{k+1},\ldots,q_{k+\ell},x_\ell)$. Conversely, if $x = (q_1,\ldots,q_k,x_k)$ and $x_k = (q_{k+1},\ldots,q_{k+\ell},x_\ell)$ with $k,\ell \geqslant 1$ then $x = (q_1,\ldots,q_k,q_{k+1},\ldots,q_{k+\ell},x_\ell)$.

Proof The second assertion is immediate, since it amounts only to carrying the continued fraction algorithm beyond k. To prove the first assertion, set

$$x_k' = (q_{k+1},\ldots,q_{k+\ell},x_\ell).$$

Since $\ell > 1$, there are at least two terms so $x_k' > 1$. Evaluating the continued fraction from below, it is clear that $x = (q_1,\ldots,q_k,x_k')$. But by Proposition 2, it follows that $(q_1,\ldots,q_k,x_k')$ is the continued fraction expansion of x; hence $x_k = x_k'$.

DEFINITION We define the infinite continued fraction by

$$(q_1,q_2,\ldots) = \lim_{k\to\infty} (q_1,q_2,\ldots,q_k) = \lim_{k\to\infty} \frac{Z_k}{N_k}.$$

PROPOSITION 3 Let the real irrational number x have the finite continued fraction $x = (q_1,q_2,\ldots,q_k,x_k)$ and the infinite continued fraction $(q_1,q_2,\ldots)$ then $x_k = (q_{k+1},q_{k+2},\ldots)$. Conversely, if $x = (q_1,\ldots,q_k,x_k)$ and $x_k = (q_{k+1},q_{k+2},\ldots)$ then $x = (q_1,q_2,\ldots,q_k,q_{k+1},\ldots)$.

Proof Let $x = (q_1,\ldots,q_k,x_k) = (q_1,q_2,\ldots)$.

By corollary 3, $x_k = (q_{k+1},\ldots,q_{k+\ell},x_\ell)$. Hence

$$x_k = \lim_{\ell\to\infty} (q_{k+1},\ldots,q_{k+\ell}) = (q_{k+1},q_{k+2},\ldots)$$ by Corollary 2 and the definition. The converse is immediate.

6.63.1

63. EQUIVALENCE OF REAL IRRATIONAL NUMBERS

We wish to define a relation of equivalence on the set of real numbers. Two real numbers x and x' are equivalent, symbolized by $x \sim x'$, if and only if

$$x' = \frac{\alpha x + \beta}{\gamma x + \delta} \ ,$$

which we will abbreviate by

$$x' = \begin{pmatrix} \alpha & \beta \\ \gamma & \delta \end{pmatrix} x \ ,$$

where $\alpha, \beta, \gamma, \delta$ are integers and

$$\begin{vmatrix} \alpha & \beta \\ \gamma & \delta \end{vmatrix} = \pm 1.$$

If $\begin{vmatrix} \alpha & \beta \\ \gamma & \delta \end{vmatrix} = +1$, then x and x' are called primitively equivalent and if $\begin{vmatrix} \alpha & \beta \\ \gamma & \delta \end{vmatrix} = -1$ they are imprimitively equivalent. The following relations are obvious

I) $x \sim x.$

II) $x \sim x' \Rightarrow x' \sim x.$

III) $x \sim x'$ and $x' \sim x'' \Rightarrow x \sim x''.$

<u>REMARK</u> It is possible for a real irrational number to be both primitively and imprimitively equivalent to another irrational number, or to itself.*

<u>PROPOSITION 1</u> A real irrational number is equivalent to each final denominator in its continued fraction decompositions;

$$x \sim x_k \text{ if } x = (q_1, q_2, \ldots, q_k, x_k) \text{ for } k > 1 \text{ with } x_k > 1.$$

<u>PROOF</u> For $k = 1$ we have

$$x = q_1 + \frac{1}{x_1} = \frac{q_1 x_1 + 1}{1 \cdot x_1 + 0} = \begin{pmatrix} q_1 & 1 \\ 1 & 0 \end{pmatrix} x_1 \text{ with } \begin{vmatrix} q_1 & 1 \\ 1 & 0 \end{vmatrix} = -1,$$

*See problems.

6.63.2

and for $k > 2$, according to the previous section,

$$x = \frac{Z_k x_k + Z_{k-1}}{N_k x_k + N_{k-1}} = \begin{pmatrix} Z_k & Z_{k-1} \\ N_k & N_{k-1} \end{pmatrix} x_k \text{ with } \begin{vmatrix} Z_k & Z_{k-1} \\ N_k & N_{k-1} \end{vmatrix} = (-1)^k.$$

<u>PROPOSITION 2</u> Suppose that the continued fractions of the irrational numbers x and x' differ only in their initial segments (which need not be the same length);

$$x = (q_1, q_2, \ldots, q_k, s_1, s_2, \ldots)$$

$$x' = (r_1, r_2, \ldots, r_h, s_1, s_2, \ldots).$$

Then $x \sim x'$.

<u>Proof</u> We set $x'' = (s_1, s_2, \ldots) = \lim_{n \to \infty} (s_1, s_2, \ldots, s_n)$.

According to proposition 3 of the last section,

$$x = (q_1, \ldots, q_k, x'') \text{ and } x' = (r_1, \ldots, r_h, x'').$$

Thus, according to Proposition 1, $x \sim x'' \sim x'$ so $x \sim x'$.

<u>COMMENT</u> The equivalence $x \sim x'$ is primitive if and only if $h+k \equiv 0 \pmod{2}$, which is equivalent to $h \equiv k \bmod 2$.

We will now show that the converse of proposition 2 holds.

<u>PROPOSITION 3</u> (THEOREM OF LAGRANGE). Let x and x' be two equivalent irrational numbers:

$$x \sim x' = \frac{\alpha x + \beta}{\gamma x + \delta} = \begin{pmatrix} \alpha & \beta \\ \gamma & \delta \end{pmatrix} x \text{ with } \begin{vmatrix} \alpha & \beta \\ \gamma & \delta \end{vmatrix} = \pm 1 = (-1)^\ell.$$

Let the continued fraction expansions be

$$x = (q_1, \ldots, q_k, x_k) \qquad k = 1, 2, \ldots$$

$$x' = (r_1, \ldots, r_h, x_h') \qquad h = 1, 2, \ldots .$$

We may then determine integers k and h such that both

1. $x_k = x_h'$

<u>and</u> 2. the relation $x' = \dfrac{\alpha x + \beta}{\gamma x + \delta}$ results from the elimination of

$x_k = x_h'$ from the relations

$$x = (q_1,\ldots,q_k,x_k) \qquad \text{and}$$

$$x' = (r_1,\ldots,r_h,x_h').$$

Additionally, we have $k+h \equiv \ell \pmod 2$.

For the proof we will use the following lemma.

<u>LEMMA</u> Let $x = \dfrac{Ay+B}{Cy+D}$, where A,B,C,D are integers, $\begin{vmatrix} A & B \\ C & D \end{vmatrix} = \varepsilon = \pm 1$, $C > D > 0$ and $y > 1$. Then

$$x = (q_1,\ldots,q_k,y) \text{ where } (q_1,\ldots,q_k) = A/C \text{ and } (-1)^k = \varepsilon.$$

In addition, $\begin{pmatrix} A & B \\ C & D \end{pmatrix} = \begin{pmatrix} Z_k & Z_{k-1} \\ N_k & N_{k-1} \end{pmatrix}$.

<u>Proof</u> From the hypothesis it follows that $(A,C) = 1$ and $D \geqslant 1$, $C \geqslant 2$.

Expanding the rational non-integral number A/C as a continued fraction, we have

$$\frac{A}{C} = (q_1,\ldots,q_k) \text{ where } k \geqslant 2,$$

and we may also require that $(-1)^k = \varepsilon$, for in the contrary case we may, as we know, expand the continued fraction one term to $(q_1,\ldots,(q_k-1),1)$.

From $A/C = Z_k/N_k$ and $(A,C) = 1$ and $(Z_k,N_k) = 1$ and $C,N_k > 0$ it follows that $A = Z_k$ and $C = N_k$. We set

$$B' = Z_{k-1} \qquad D' = N_{k-1}.$$

Then we will have $D' > 0$ and

$$AD'-B'C = Z_k N_{k-1} - Z_{k-1} N_k = (-1)^k = \varepsilon,$$

so $AD'-B'C = AD-BC$.

Consequently $A(D'-D) = C(B'-B)$.

From $0 < D < C$ and $0 < D' = N_{k-1} < N_k = C$ we have

$$|D'-D| < C.$$

6.63.4

However $C|A(D'-D)$ and since $(A,C) = 1$ we have $C|D'-D$. But from $|D'-D| < C$ and $C|D'-D$ we get $D' = D$ and then $B' = B$. Consequently

$$(q_1,\ldots,q_k,y) = \frac{Z_k y + Z_{k-1}}{N_k y + N_{k-1}} = \frac{Ay+B}{Cy+D} = x.$$

<u>Proof of the Theorem of Lagrange.</u> We have $x = (q_1,\ldots,q_k,x_k)$ with $x_k > 1$ and irrational for $k = 1,2,3,\ldots$; and for $k \geqslant 2$

$$x = \frac{Z_k x_k + Z_{k-1}}{N_k x_k + N_{k-1}} = \begin{pmatrix} Z_k & Z_{k-1} \\ N_k & N_{k-1} \end{pmatrix} x_k \text{ with } \begin{vmatrix} Z_k & Z_{k-1} \\ N_k & N_{k-1} \end{vmatrix} = (-1)^k.$$

From the hypothesis

$$x' = \frac{\alpha x + \beta}{\gamma x + \delta} \quad \text{with} \quad \begin{vmatrix} \alpha & \beta \\ \gamma & \delta \end{vmatrix} = (-1)^\ell,$$

so

$$x' = \begin{pmatrix} \alpha & \beta \\ \gamma & \delta \end{pmatrix}\begin{pmatrix} Z_k & Z_{k-1} \\ N_k & N_{k-1} \end{pmatrix} x_k \text{ for } k = 2,3,\ldots \ ,$$

and thus

$$x' = \begin{pmatrix} A_k & B_k \\ C_k & D_k \end{pmatrix} x_k \text{ with } \begin{pmatrix} A_k & B_k \\ C_k & D_k \end{pmatrix} = \begin{pmatrix} \alpha & \beta \\ \gamma & \delta \end{pmatrix}\begin{pmatrix} Z_k & Z_{k-1} \\ N_k & N_{k-1} \end{pmatrix}.$$

We may assume that $\gamma x + \delta > 0$ because, in the contrary case, instead of

$$x' = \frac{\alpha x + \beta}{\gamma x + \delta} , \tag{63.1}$$

we may take

$$x' = \frac{-\alpha x - \beta}{-\gamma x - \delta} \text{ with } \begin{vmatrix} -\alpha & -\beta \\ -\gamma & -\delta \end{vmatrix} = \begin{vmatrix} \alpha & \beta \\ \gamma & \delta \end{vmatrix} = (-1)^\ell.$$

We have

$$C_k = N_k \left(\gamma \frac{Z_k}{N_k} + \delta \right), \ D_k = N_{k-1}\left(\gamma \frac{Z_{k-1}}{N_{k-1}} + \delta \right) \text{ for } k = 2,3,\ldots \ .$$

Since

$$\lim_{k\to\infty} \frac{Z_k}{N_k} = \lim_{k\to\infty} \frac{Z_{k-1}}{N_{k-1}} = x \text{ and } \gamma x + \delta > 0 ,$$

we will have, for sufficiently large k,

$$\gamma \frac{Z_k}{N_k} + \delta > 0 \text{ and } \gamma \frac{Z_{k-1}}{N_{k-1}} + \delta > 0,$$

so

$$C_k > 0, \ D_k > 0.$$

6.63.5

Moreover,

$$\frac{D_k}{C_k} = \frac{\gamma Z_{k-1} + \delta N_{k-1}}{\gamma Z_k + \delta N_k} = \frac{N_{k-1}}{N_k} + \left(\frac{\gamma Z_{k-1} + \delta N_{k-1}}{\gamma Z_k + \delta N_k} - \frac{N_{k-1}}{N_k}\right)$$

$$= \frac{N_{k-1}}{N_k} + \frac{\gamma(Z_{k-1}N_k - Z_k N_{k-1})}{N_k(\gamma Z_k + \delta N_k)}$$

$$= \frac{N_{k-1}}{N_k} + \frac{\pm\gamma}{N_k(\gamma Z_k + \delta N_k)}$$

$$= \frac{N_{k-1}}{N_k} + \frac{\theta_k}{N_k} \quad \text{with} \quad \theta_k = \frac{\pm\gamma}{\gamma Z_k + \delta N_k} \; .$$

However,

$$\lim_{k\to\infty} \theta_k = \lim_{k\to\infty} \frac{1}{N_k} \frac{\pm\gamma}{\gamma(Z_k/N_k) + \delta} = 0$$

because, since $\lim_{k\to\infty} Z_k/N_k = x$ and x is irrational, $\gamma(Z_k/N_k) + \delta$ cannot

approach 0. Hence, for sufficiently large k, $|\theta_k| < 1$ and thus

$$\frac{D_k}{C_k} = \frac{N_{k-1} + \theta_k}{N_k} < 1, \text{ so } 0 < D_k < C_k.$$

We have found that for sufficiently large k

$$x' = \begin{pmatrix} A_k & B_k \\ C_k & D_k \end{pmatrix} x_k \quad \text{with} \quad C_k > D_k > 0,$$

and

$$\begin{vmatrix} A_k & B_k \\ C_k & D_k \end{vmatrix} = \begin{vmatrix} \alpha & \beta \\ \gamma & \delta \end{vmatrix} \begin{vmatrix} Z_k & Z_{k-1} \\ N_k & N_{k-1} \end{vmatrix} = (-1)^{\ell+k}.$$

We may now apply the lemma, putting x' for x, x_k for y and

$\begin{pmatrix} A_k & B_k \\ C_k & D_k \end{pmatrix}$ for $\begin{pmatrix} A & B \\ C & D \end{pmatrix}$. We obtain

$$x' = (s_1, s_2, \ldots, s_h, x_k),$$

where

$$(s_1, \ldots, s_h) = A_k/C_k \text{ and } (-1)^h = (-1)^{\ell+k}.$$

First, we have $h \equiv \ell+k \pmod 2$ so $h+k \equiv \ell \pmod 2$.

6.63.6

Also, we have
$$\begin{pmatrix} A_k & B_k \\ C_k & D_k \end{pmatrix} = \begin{pmatrix} Z'_h & Z'_{h-1} \\ N'_h & N'_{h-1} \end{pmatrix}$$
where Z'_i and N'_i are the partial quotients for $(s_1,\ldots,s_h,x_k)$.

Now expand x' in a continued fraction of length h, $x' = (r_1,r_2,\ldots,r_h,x'_h)$. Since we also have $x' = (s_1,s_2,\ldots,s_h,x_k)$ where x_k, being a final denominator for x, satisfies $x_k > 1$. It follows by Proposition 2 of §62 that $r_i = s_i$, $i = 1,\ldots,h$ and $x_k = x'_h$ and

$$x = (q_1,\ldots,q_k,x_k)$$
$$x' = (r_1,\ldots,r_h,x_k),$$

as required.

Finally, since
$$\begin{pmatrix} \alpha & \beta \\ \gamma & \delta \end{pmatrix}\begin{pmatrix} Z_k & Z_{k-1} \\ N_k & N_{k-1} \end{pmatrix} = \begin{pmatrix} Z'_h & Z'_{h-1} \\ N'_h & N'_{h-1} \end{pmatrix}, \tag{63.2}$$
we have
$$(-1)^k \begin{pmatrix} \alpha & \beta \\ \gamma & \delta \end{pmatrix} = \begin{pmatrix} Z'_h & Z'_{h-1} \\ N'_h & N'_{h-1} \end{pmatrix}\begin{pmatrix} N_{k-1} & -Z_{k-1} \\ -N_k & Z_k \end{pmatrix}. \tag{63.3}$$

Thus*
$$x' = \begin{pmatrix} \alpha & \beta \\ \gamma & \delta \end{pmatrix}x = \begin{pmatrix} Z'_h & Z'_{h-1} \\ N'_h & N'_{h-1} \end{pmatrix}\begin{pmatrix} N_{k-1} & -Z_{k-1} \\ -N_k & Z_k \end{pmatrix}x.$$

This is the matrix form for eliminating x_k from
$$x = (q_1,\ldots,q_k,x_k) = \begin{pmatrix} Z_k & Z_{k-1} \\ N_k & N_{k-1} \end{pmatrix}x_k$$
$$x' = (r_1,\ldots,r_h,x_k) = \begin{pmatrix} Z'_h & Z'_{h-1} \\ N'_h & N'_{h-1} \end{pmatrix}x_k,$$

and the theorem has been completely proved.

<u>COROLLARY</u> With the same assumptions and notation of the last proposition,
$$\begin{pmatrix} \alpha & \beta \\ \gamma & \delta \end{pmatrix} = \begin{pmatrix} Z'_h & Z'_{h-1} \\ N'_h & N'_{h-1} \end{pmatrix}\begin{pmatrix} Z_k & Z_{k-1} \\ N_k & N_{k-1} \end{pmatrix}^{-1}.$$
where we assume, as in (63.1), that $\gamma x + \delta > 0$.

*Note $(-S)x = (-\alpha x - \beta)/(-\gamma x - \delta) = (\alpha x + \beta)/(\gamma x + \delta) = Sx$

6.63.7

<u>Proof</u> This is simply (63.2) in the proposition.

<u>REMARK</u> In the theorem of Lagrange, once k and h have been determined

satisfying 1) and 2) of the theorem, the numbers $k+\ell$ and $h+\ell$ (ℓ a

non-negative integer) will also satisfy 1) and 2). Indeed, the

continued fractions will be identical beyond position k for x and

position h for x'.

64. REDUCED QUADRATIC FORMS WITH DISCRIMINANT $\Delta = b^2 - ac$ POSITIVE

Let $f = (a,b,c) = ax^2 + 2bxy + cy^2$ be a quadratic form with discriminant $\Delta = b^2 - ac > 0$ where Δ is not the square of an integer. Then $ac \neq 0$ so $a \neq 0$ and $c \neq 0$. The roots ω and η of the form f will be real irrational numbers. We have

$$\omega = \frac{-b-\sqrt{\Delta}}{a} = \frac{(-b-\sqrt{\Delta})(-b+\sqrt{\Delta})}{a(-b+\sqrt{\Delta})} = \frac{b^2-\Delta}{a(-b+\sqrt{\Delta})} = \frac{c}{-b+\sqrt{\Delta}}$$

$$\eta = \frac{-b+\sqrt{\Delta}}{a} = \frac{(-b+\sqrt{\Delta})(-b-\sqrt{\Delta})}{a(-b-\sqrt{\Delta})} = \frac{b^2-\Delta}{a(-b-\sqrt{\Delta})} = \frac{c}{-b-\sqrt{\Delta}}$$

and

$$\eta - \omega = \frac{2\sqrt{\Delta}}{a} \ .$$

<u>PROPOSITION 1</u> A quadratic form $f = (a,b,c)$ with $\Delta > 0$ and not the square of an integer is completely determined by Δ and ω.

<u>Proof</u> Indeed, let $f = (a,b,c)$ and $f_1 = (a_1,b_1,c_1)$ be two quadratic forms with the same non-square $\Delta > 0$ and the same ω. Then we would have

$$\frac{-b-\sqrt{\Delta}}{a} = \frac{-b_1 - \sqrt{\Delta}}{a_1} \quad \text{where } \sqrt{\Delta} \text{ is irrational.}$$

So

$$\frac{-b}{a} = \frac{-b_1}{a_1} \quad \text{and} \quad \frac{-1}{a} = \frac{-1}{a_1} \ .$$

From these two equations follows first $a = a_1$ and then $b = b_1$. Then from $b^2 - ac = \Delta = b_1^2 - a_1 c_1$ follows $c = c_1$, since $a \neq 0$.

<u>DEFINITION</u> The quadratic form $f = (a,b,c)$ with $\Delta > 0$ is called reduced when

$$|\omega| > 1, \ |\eta| < 1, \ \omega\eta < 0.$$

6.64.2

<u>PROPOSITION 2</u> If $f = (a,b,c)$ with $\Delta > 0$ is reduced, then

$ac < 0$ and $|b| < \sqrt{\Delta}$.

<u>Proof</u> From $\omega n = c/a < 0$ follows $ac < 0$. Moreover, from $\Delta = b^2 - ac$ and

$ac < 0$ follow $\Delta > b^2$ and thus $\sqrt{\Delta} > |b|$.

<u>CAUTION</u> The converse of Proposition 2 is not true. For example

$x^2 - 2xy - y^2$ has $ac < 0$, $b^2 = (-1)^2 < \Delta = 2$ but

$$|\omega| = |\tfrac{1-\sqrt{2}}{1}| < 1 \text{ and } |n| = |\tfrac{1+\sqrt{2}}{1}| > 1,$$

so $x^2 - 2xy - y^2$ is not reduced.

<u>PROPOSITION 3</u> A quadratic form $f = (a,b,c)$ with $\Delta > 0$ is reduced if and

only if

$$\sqrt{\Delta} + b > |a| > \sqrt{\Delta} - b > 0. \qquad\qquad (64.1)$$

<u>Proof</u> Let first $f = (a,b,c)$ be reduced; then, according to

Proposition 2,

$$|b| < \sqrt{\Delta} \text{ and thus } \pm b + \sqrt{\Delta} > 0.$$

From this we have $|-b-\sqrt{\Delta}| = b + \sqrt{\Delta}$ so

$$1 < |\omega| = |\tfrac{-b-\sqrt{\Delta}}{a}| = \tfrac{b+\sqrt{\Delta}}{|a|} , \quad \text{ so } b + \sqrt{\Delta} > |a|,$$

and also $|-b+\sqrt{\Delta}| = -b + \sqrt{\Delta}$ so

$$1 > |n| = |\tfrac{-b+\sqrt{\Delta}}{a}| = \tfrac{-b+\sqrt{\Delta}}{|a|} , \quad \text{ so } -b + \sqrt{\Delta} < |a|,$$

as required.

For the converse, suppose (64.1) holds. From this obviously follows

$$|\omega| = \tfrac{|-b-\sqrt{\Delta}|}{|a|} = \tfrac{|b+\sqrt{\Delta}|}{|a|} = \tfrac{b+\sqrt{\Delta}}{|a|} > 1$$

and

$$|\eta| = \frac{|-b+\sqrt{\Delta}|}{|a|} = \frac{-b+\sqrt{\Delta}}{|a|} < 1$$

and finally

$$\omega\eta = \frac{b^2-\Delta}{a^2} < 0,$$

because $\sqrt{\Delta}+b > \sqrt{\Delta}-b > 0$ implies

$$\Delta-b^2 = (\sqrt{\Delta}+b)(\sqrt{\Delta}-b) > 0,$$

so $b^2-\Delta < 0$.

PROPOSITION 4 In a reduced form $f = (a,b,c)$ with $\Delta > 0$ we have

$$b > 0, \quad -a\omega > 0, \quad c\omega > 0.$$

Proof From $\sqrt{\Delta}+b > \sqrt{\Delta}-b$ we get $2b > 0$ so $b > 0$. Moreover $a\omega = -b-\sqrt{\Delta}$
$= -(b+\sqrt{\Delta}) < 0$ because $b+\sqrt{\Delta} > 0$. Finally, since a and c have opposite
sign (recall $\frac{c}{a} = \omega\eta < 0$ in a reduced form) we have $c\omega > 0$.

EXAMPLES The forms $(1,1,-1)$ and $(-1,1,1)$ are reduced. Also reduced are
$(2,1,-1)$ and $(-2,1,1)$.

REMARK If (a,b,c) with $\Delta > 0$ is reduced, the $(-a,b,-c)$ will also be
reduced. Actually, the condition

$$\sqrt{\Delta}+b > |a| > \sqrt{\Delta}-b > 0$$

implies

$$\sqrt{\Delta}+b > |-a| > \sqrt{\Delta}-b > 0.$$

PROPOSITION 5 To a given discriminant $\Delta > 0$ correspond only finitely
many reduced quadratic forms.

Proof Since $ac < 0$ we have $\Delta = b^2-ac = b^2+|ac|$ so $b^2 < \Delta$, and b can have
only the values $1,2,\ldots,[\sqrt{\Delta}]$. (The value $b = 0$ is excluded by
proposition 4.) Since $-ac = \Delta-b^2$, and for each value of b there are
only finitely many ways to represent $\Delta-b^2$ as a product of two

6.64.4

integers, the total number of forms (a,b,c) with discriminant Δ is finite.

<u>PROPOSITION 6</u> Each quadratic form with positive discriminant not the square of an integer is equivalent to a reduced form.

<u>Proof</u> Let ω be the first root of the quadratic form f (see §56). We expand ω, which is an irrational number, in a continued fraction:

$$\omega = (q_1, q_2, \ldots, q_k, \omega_k) \quad \text{for } k = 1, 2, 3, \ldots . \tag{64.2}$$

We take k even; then

$$\omega = \frac{Z_k \omega_k + Z_{k-1}}{N_k \omega_k + N_{k-1}} \qquad \text{where } \begin{vmatrix} Z_k & Z_{k-1} \\ N_k & N_{k-1} \end{vmatrix} = (-1)^k = 1.$$

We designate $\begin{pmatrix} Z_k & Z_{k-1} \\ N_k & N_{k-1} \end{pmatrix}$ by S_k. According to proposition 3 of §56, if we call ω_k and η_k the first and second roots respectively of $\phi_k = fS_k$ we will have

$$\omega = S_k \omega_k = \frac{Z_k \omega_k + Z_{k-1}}{N_k \omega_k + N_{k-1}} \qquad \eta = S_k \eta_k = \frac{Z_k \eta_k + Z_{k-1}}{N_k \eta_k + N_{k-1}} .$$

We have $\omega_k > 1$ for $k > 1$ since ω_k is the final denominator of the continued fraction expansion (64.2). I say that for sufficiently large k the form ϕ_k is reduced. Actually, we have already that $\omega_k > 1$ and we will now prove that, for k sufficiently large, $-1 < \eta_k < 0$.

We have

$$\eta_k = S_k^{-1}\eta = \frac{N_{k-1}\eta - Z_{k-1}}{-N_k \eta + Z_k} = - \frac{N_{k-1} + \theta_k}{N_k} = - \frac{N_{k-1}}{N_k} - \frac{\theta_k}{N_k}$$

where

$$\frac{\theta_k}{N_k} = \frac{N_{k-1}\eta - Z_{k-1}}{N_k \eta - Z_k} - \frac{N_{k-1}}{N_k}$$

so

$$\theta_k = \frac{N_k(N_{k-1}\eta - Z_{k-1})}{N_k \eta - Z_k} - N_{k-1}$$

6.64.5

$$= \frac{Z_k N_{k-1} - Z_{k-1} N_k}{N_k n - Z_k}$$

$$= \frac{1}{N_k \left(n - \frac{Z_k}{N_k}\right)} \qquad \text{(since k is even)}.$$

However, $\lim_{k \to \infty} \dfrac{Z_k}{N_k} = \omega$ and thus

$$\lim_{k \to \infty} \left(n - \frac{Z_k}{N_k}\right) = n - \omega \neq 0.$$

Consequently,

$$\lim_{k \to \infty} \theta_k = 0, \qquad \text{since } \lim_{k \to \infty} N_k = \infty.$$

Thus for sufficiently large k we have $-1 < \theta_k < 1$ and then

$$-\frac{N_{k-1}}{N_k} - \frac{1}{N_k} < \eta_k < -\frac{N_{k-1}}{N_k} + \frac{1}{N_k} .$$

Since $-N_{k-1}+1 < 0$ and $N_{k-1}+1 < N_k$ for $k \geqslant 5$ (since $N_k > k-1$) we have, for

sufficiently large k

$$-1 < \eta_k < 0,$$

as required.

From proposition 5 and 6 we may derive the important:

PROPOSITION 7 The first root of a quadratic form with positive

discriminant Δ not the square of an integer has a periodic

continued fraction

$$\omega = (q_1, \ldots, q_k, \overline{r_1, \ldots, r_j});$$

(where the bar over $r_1, \ldots, r_j$ indicates that it is repeated endlessly).

6.64.6

Proof As we saw in the proof of Proposition 6, from a certain k on the forms $\phi_k = fS_k$ (k even) are reduced. However, the number of reduced positive forms with a given discriminant is finite (Proposition 5), so there exist even integers k and ℓ, k $<$ ℓ, for which $\phi_k = \phi_\ell$. Then we will have $\omega_k = \omega_\ell$. Consequently,

$$\omega = (q_1,\ldots,q_k,\omega_k) = (q_1,\ldots,q_k,q_{k+1},\ldots,q_\ell,\omega_\ell)$$
$$= (q_1,\ldots,q_k,q_{k+1},\ldots,q_\ell,\omega_k),$$

and thus $\omega_k = (q_{k+1},\ldots,q_\ell,\omega_k)$. Setting $r_1 = q_{k+1},\ldots,r_j = q_{k+j}$ with k+j = ℓ, we have

$$\omega = (q_1,\ldots,q_k,r_1,\ldots,r_j,\omega_k) = (q_1,\ldots,q_k,r_1,\ldots,r_j,r_1,\ldots,r_j,\omega_k)$$
$$= (q_1,\ldots,q_k,\overline{r_1,\ldots,r_j}).$$

COROLLARY A real irrational number of the form $\sqrt{m/n}$ (m and n relatively prime positive integers) expands into a periodic continued fraction:
$$\sqrt{\frac{m}{n}} = (q_1,\ldots,q_k,\overline{r_1,\ldots,r_j}).$$

Proof $\sqrt{\dfrac{m}{n}} = \dfrac{\sqrt{mn}}{n}$. The quadratic form $-nx^2+my^2$ has for its first root $\omega = \dfrac{-0-\sqrt{mn}}{-n} = \dfrac{\sqrt{mn}}{n} = \sqrt{\dfrac{m}{n}}$, and by the last Proposition ω has a periodic continued fraction.

PROBLEM Given the quadratic forms f and f´ with the same positive discriminant not the square of an integer; we ask, is f $\sim$ f´?

To answer, we find two reduced forms ϕ and $\phi´$ equivalent to f and f´ respectively:

$$fS = \phi \quad \text{and} \quad f´S´ = \phi´.$$

Then f $\sim$ f´ if and only if $\phi \sim \phi´$.

The problem of the equivalence of reduced forms will be solved in the next section.

65. THE PERIODS OF A REDUCED QUADRATIC FORM WITH $\Delta > 0$

Let $f = (a,b,a')$ be reduced with positive discriminant Δ not the square of an integer. We ask: Is there a right adjacent form $f_1 = (a',b',a'')$ which is also reduced?

We have, if the answer is affirmative,

$$f_1 = (a,b,a')\begin{pmatrix} 0 & -1 \\ 1 & k \end{pmatrix} = (a',b',a'').\tag{65.1}$$

Since f is reduced, we will have, in accordance with Propositions 2 and 4 of the preceeding section, the relations $a'\omega > 0$, $a'\eta < 0$. Since f_1 is, by hypothesis, also reduced, we will have $a'\omega_1 < 0$, $a'\eta_1 > 0$, where ω_1 and η_1 are the first and second roots of f_1. Let $\mathrm{sgn}(a') = \varepsilon,$* from which we get $\mathrm{sgn}(\eta) = -\varepsilon$, $\mathrm{sgn}(\omega) = \varepsilon$, $\mathrm{sgn}(\eta_1) = \varepsilon$ and $\mathrm{sgn}(\omega_1) = -\varepsilon$. Because of (65.1) we now have the relations

$$\omega = \frac{-1}{\omega_1 + k} \qquad \eta = \frac{-1}{\eta_1 + k},$$

and thus

$$\omega_1 = \frac{k\omega + 1}{-\omega} = -k - \frac{1}{\omega} \qquad \frac{1}{\eta} = -k - \eta_1,$$

from which we conclude

$$|\omega_1| = -\varepsilon\omega_1 = \varepsilon k + \frac{1}{\varepsilon\omega} \qquad \left|\frac{1}{\eta}\right| = \frac{-\varepsilon}{\eta} = k\varepsilon + \varepsilon\eta_1 = k\varepsilon + \frac{1}{|1/\eta_1|}.$$

---

*$\mathrm{sgn}(x)$ is defined by $\mathrm{sgn}(x) = \begin{cases} +1 & \text{if } x > 0 \\ 0 & \text{if } x = 0 \\ -1 & \text{if } x < 0 \end{cases}.$

6.65.2

From $\mathrm{sgn}(\eta_1) = \varepsilon$ if follows that $\varepsilon\eta_1 > 0$; however $|\eta_1| < 1$, so we have $0 < \varepsilon\eta_1 < 1$. From this and $-\dfrac{\varepsilon}{\eta} = \varepsilon k + \varepsilon\eta_1$ we conclude that $\varepsilon k = [-\varepsilon/\eta]$. Thus, if there exists a right adjacent reduced form, k is uniquely determined by $k = \varepsilon[-\varepsilon/\eta]$.

Conversely, let us take $k = \varepsilon[-\varepsilon/\eta]$, from which follows

$$\varepsilon k = \left[\frac{-\varepsilon}{\eta}\right].$$

Since $\mathrm{sgn}(\eta) = -\varepsilon$ and $|\eta| < 1$, we will have $-\varepsilon/\eta > 1$ and consequently $\varepsilon k \geq 1$. We now set

$$\varepsilon\eta_1 = \frac{-\varepsilon}{\eta} - \varepsilon k = \frac{-\varepsilon}{\eta} - \left[\frac{-\varepsilon}{\eta}\right] ,$$

so

$$0 < \varepsilon\eta_1 < 1.$$

(It is not possible that $-\dfrac{\varepsilon}{\eta} = \left[-\dfrac{\varepsilon}{\eta}\right]$, since $\dfrac{-\varepsilon}{\eta}$ is irrational).

From $\varepsilon\eta_1 > 0$ it follows that $\mathrm{sgn}(\eta_1) = \varepsilon$ and from $\mathrm{sgn}(a') = \varepsilon$ then it follows that $a'\eta_1 > 0$.

We also set

$$-\varepsilon\omega_1 = \varepsilon k + \frac{1}{\varepsilon\omega} ,$$

from which we get $-\varepsilon\omega_1 \geq 1+(1/\varepsilon\omega)$. But $\varepsilon\omega > 0$ since $\mathrm{sgn}\,\omega = \varepsilon$, so $-\varepsilon\omega_1 > 1$, which implies $\mathrm{sgn}(\omega_1) = -\varepsilon$. Finally, from $0 < \varepsilon\eta_1 < 1$ follows $|\eta_1| < 1$ and from $-\varepsilon\omega_1 > 1$ follows $|\omega_1| > 1$, while $\omega_1\eta_1 = -(-\varepsilon\omega_1)(\varepsilon\eta_1) < 0$. But the ω_1 and η_1 defined above satisfy

$$\omega = \frac{-1}{\omega_1+k} \quad \text{and} \quad \eta = \frac{-1}{\eta_1+k} ,$$

so ω_1 and η_1 are the first and second roots respectively of $(a,b,a')\begin{pmatrix} 0 & -1 \\ 1 & k \end{pmatrix}$. Thus $(a,b,a')\begin{pmatrix} 0 & -1 \\ 1 & k \end{pmatrix}$, with the above defined $k = \varepsilon[-\varepsilon/\eta]$, is reduced. Thus there is exactly one right adjacent reduced form to (a,b,a') and we have proved:

PROPOSITION 1 Each reduced quadratic form $f = (a,b,a')$ with positive discriminant not the square of an integer has exactly one right adjacent reduced form $f_1 = (a',b',a'')$. The first roots ω and ω_1 of f and f_1 satisfy

$$\text{sgn}(\omega_1) = -\text{sgn}(\omega).$$

We also have the relations

$$|\omega_1| = p+\frac{1}{|\omega|} \qquad\qquad |\frac{1}{\eta}| = p+\frac{1}{|1/\eta_1|} \qquad\qquad (65.2)$$

where $p = \epsilon k$ is a positive integer.

DEFINITION The form f is left adjacent to the form f_1 if and only if f_1 is right adjacent to f.

PROPOSITION 2 Each reduced quadratic form $f = (a',b',a'')$ with positive discriminant not the square of an integer has exactly one left adjacent reduced form $f_2 = (a,b,a')$.

Proof* We must have, for a suitable k,

$$(a,b,a')\begin{pmatrix}0 & -1\\ 1 & k\end{pmatrix} = (a',b',a''),$$

which is equivalent to

$$(a,b,a') = (a',b',a'')\begin{pmatrix}k & 1\\ -1 & 0\end{pmatrix}.$$

Set $\epsilon' = \text{sgn}(a')$. Let ω and η be the roots of the given $f = (a',b',a'')$ and ω_2,η_2 the roots of $f_2 = (a,b,a')$. Then, since $-a'\omega > 0$ (Proposition 4 of §64), we have $\text{sgn}(\omega) = -\epsilon'$ and then, since $\omega\eta < 0$, $\text{sgn}(\eta) = \epsilon'$.

Let us first assume that f_2 is also reduced. Then, as above, $a'\omega_2 > 0$ so $\text{sgn}(\omega_2) = \text{sgn}(a') = \epsilon'$ and $\text{sgn}(\eta_2) = -\epsilon'$. Moreover

$$\omega = \begin{pmatrix}k & 1\\ -1 & 0\end{pmatrix}\omega_2 \qquad \eta = \begin{pmatrix}k & 1\\ -1 & 0\end{pmatrix}\eta_2,$$

so

---

*A second proof, which depends on proposition 1, is given in the problems.

6.65.4

$$\omega = -k-\frac{1}{\omega_2} \qquad \frac{1}{\eta_2} = -k-\eta \ ,$$

and consequently

$$-\varepsilon'\omega = k\varepsilon'+\frac{1}{\varepsilon'\omega_2} \qquad \frac{-\varepsilon'}{\eta_2} = k\varepsilon'+\varepsilon'\eta.$$

Since $\varepsilon'\omega_2 > 0$ and $\varepsilon'\omega_2 = |\omega_2| > 1$, we have

$$k\varepsilon' = [-\varepsilon'\omega] \quad \text{so} \quad k = \varepsilon'[-\varepsilon'\omega].$$

The number k is thus completely determined; there is at most one left adjacent reduced form.

Conversely, the k defined by $k = \varepsilon'[-\varepsilon'\omega]$ does indeed produce a left adjacent reduced form. Note first that $-\varepsilon'\omega = |\omega| > 1$ so $k\varepsilon' \geqslant 1$. Define ω_2 by

$$-\varepsilon'\omega = k\varepsilon' + \frac{1}{\varepsilon'\omega_2} \quad \text{so} \quad 0 < \frac{1}{\varepsilon'\omega_2} < 1.$$

From this follows $\varepsilon'\omega_2 > 1$, so $|\omega_2| > 1$. Next define η_2 by

$$\frac{-\varepsilon'}{\eta_2} = k\varepsilon'+\varepsilon'\eta.$$

Then we have, since $k\varepsilon' \geqslant 1$ and $\varepsilon'\eta > 0$

$$\frac{-\varepsilon'}{\eta_2} > 1 \quad \text{so} \quad |\eta_2| < 1.$$

Finally $\omega_2\eta_2 = -(\varepsilon'\omega_2)(-\varepsilon'\eta_2) < 0$. But ω_2 and η_2 as defined above are the roots of f_2. Therefore f_2 is reduced, and the theorem is proved.

We note that, setting $p = k\varepsilon'$, we have

$$|\omega| = p + \frac{1}{|\omega_2|} \quad \text{and} \quad \left|\frac{1}{\eta_2}\right| = p + \frac{1}{|1/\eta|} \tag{65.3}$$

Let now f_0 be a reduced form with positive discriminant not the square of an integer. We form successively the right adjacent reduced forms of f_0;

$$f_0 \sim f_1 \sim f_2 \sim \ldots \sim f_h \sim \ldots \sim f_k \sim \ldots \ .$$

Since there exist only finitely many reduced quadratic forms with a given discriminant Δ, there will exist an f_k identical with some previous f_h.

6.65.5

The case k = 1 is not possible, since $\text{sgn}(\omega_1) = -\text{sgn}(\omega_0)$. By the last Proposition, we will have

$$f_h = f_k \Rightarrow f_{h-1} = f_{k-1} \Rightarrow f_{h-2} = f_{k-2} \Rightarrow f_0 = f_{h-h} = f_{k-h}.$$

Then $k-h \equiv 0 \pmod 2$, for, from $f_0 = f_{k-h}$ it follows that $\omega_{k-h} = \omega_0$. However

$$\text{sgn}(\omega_0) = -\text{sgn}(\omega_1) = +\text{sgn}(\omega_2) = -\text{sgn}(\omega_3) = \ldots ,$$

so $\omega_{k-h} = \omega_0$ implies $k-h \equiv 0 \pmod 2$. If now we set $k-h = 2n$, we will have

$$f_0 \sim f_1 \sim f_2 \sim \ldots \sim f_{2n} = f_0,$$

where f_{2n} is the first of the sequence $f_1, f_2, \ldots$ which is equal to f_0.

The system of reduced quadratic forms

$$\{f_0, f_1, \ldots, f_{2n-1}\} \text{ with } f_{2n} = f_0$$

is called the period of the reduced quadratic form f_0.

Because of the uniqueness of left and right adjacent reduced forms, a quadratic form f_i of the period of f_0 will have for its period the system

$$\{f_i, f_{i+1}, \ldots, f_{2n-1}, f_0, f_1, \ldots, f_{i-1}, \},$$

which will be only a cyclic rearrangement of the period of f_0 and is not to be considered essentially different from it.

If we designate $|\omega_i|$ by y_i and $|1/\eta_i|$ by x_i which are both greater than 1, we will have, according to the above and (65.3)

$$x_0 = p_1 + \frac{1}{x_1}, \ x_1 = p_2 + \frac{1}{x_2}, \ldots, x_{2n-1} = p_{2n} + \frac{1}{x_0}$$

$$y_1 = p_1 + \frac{1}{y_0}, \ y_2 = p_2 + \frac{1}{y_1}, \ldots, y_0 = p_{2n} + \frac{1}{y_{2n-1}}$$

where the $p_j = (j = 1, 2, \ldots, 2n)$ are positive integers. Thus we find

6.65.6

$$\frac{1}{|\eta_0|} = x_0 = (p_1,p_2,\ldots,p_{2n},x_0) = \overline{(p_1,p_2,\ldots,p_{2n})}$$

$$|\omega_0| = y_0 = (p_{2n},p_{2n-1},\ldots,p_1,y_0) = \overline{(p_{2n},p_{2n-1},\ldots,p_1)}.$$

We have proved:

PROPOSITION 3 For a reduced quadratic form with positive discriminant not the square of an integer, the absolute value of the first root and the reciprocal of the absolute value of the second root expand into purely periodic continued fractions.

PROPOSITION 4 Two equivalent reduced quadratic forms f and f´ (with positive discriminant not the square of an integer) belong to the same period.

Proof Let f belong to the period

$$f_0,f_1,\ldots,f,\ldots,f_{2n-1} \tag{65.4}$$

and f´ belong to the period

$$f_0´,f_1´,\ldots,f´,\ldots,f_{2m-1}´. \tag{65.5}$$

We must prove that the two periods do not essentially differ. We designate by $\omega_k(\omega_k´)$ the first root of $f_k(f_k´)$, $(k = 0,1,2,\ldots)$. It is possible to assume that $\omega_0 > 0$ $(\omega_0´ > 0)$, for in the contrary case we may replace the sequence by a cyclic permutation of it beginning with the left or right adjacent form to $f_0(f_0´)$. According to that which we have found above

$$\omega_0 = (p_{2n},|\omega_{2n-1}|) = (p_{2n},p_{2n-1},\omega_{2n-2}) = (p_{2n},p_{2n-1},p_{2n-2},|\omega_{2n-3}|) = \ldots$$

and

$$\omega_0´ = (q_{2m},|\omega_{2m-1}´|) = (q_{2m},q_{2m-1},\omega_{2m-2}´) = (q_{2m},q_{2m-1},q_{2m-2},|\omega_{2n-3}´|) = \ldots,$$

which may be expressed

$$\omega_0 = (r_1,r_2,\ldots,r_k,|\omega_r|) \qquad k = 1,2,\ldots \tag{65.6}$$

where r is the least non-negative residue of 2n-k (mod 2n), and

$$\omega_0' = (s_1, s_2, \ldots, s_\ell, |\omega_s'|) \qquad \ell = 1, 2, \ldots, \tag{65.7}$$

where s is the least non-negative residue of 2m-ℓ (mod 2m).

Moreover, the hypothesis $f_0' \sim f' \sim f \sim f_0$ implies that $f_0' \sim f_0$ so $\omega_0 = S\omega_0' = (\alpha\omega_0' + \beta)/(\gamma\omega_0' + \delta)$ with $\alpha\delta - \beta\gamma = 1$. Consequently, according to Proposition 3 of §63 (Theorem of Lagrange), we may determine r and s such that $|\omega_r| = |\omega_s'|$ and $r \equiv s$ (mod 2). But since the ω_r and ω_s' alternate signs and $\omega_0, \omega_0' > 0$, the conditions $|\omega_r| = |\omega_s'|$ and $r \equiv s$ (mod 2) imply $\omega_r = \omega_s'$. Thus the two forms f_r and f_s' have the same determinant Δ and the same first root $\omega_r = \omega_s'$, and thus $f_r = f_s'$. Hence f_s' occurs in the period of f_0, and this must also be true of f_0', and so the two periods contain the same forms, which proves the theorem.

Now let f and f' be two quadratic forms with the same positive discriminant Δ not the square of an integer. Let $fS = \phi$ and $fS' = \phi'$ where ϕ and ϕ' are reduced. We form the period of ϕ:

$$\phi, \phi_1, \phi_2, \ldots \; .$$

According to Proposition 4, $\phi' \sim \phi$ if and only if ϕ' belongs to the period of ϕ. If this occurs, then $\phi S'' = \phi'$ and thus

$$fSS'' = \phi S'' = \phi' = f'S'$$

so

$$fSS''(S')^{-1} = f'.$$

Thus we have completely solved the first problem of §53, to which the first basic problem of §46 had been reduced.

6.65.8

EXAMPLE Using the method of Proposition 5 of §64, we may find all the reduced forms with a given discriminant and arrange them in periods. We will carry this out for $\Delta = 17$ and list the results for some other values of Δ.

Δ	PERIODS	NUMBER OF CLASSES
2	$(1,1,-1)(-1,1,1)$	1
3	$(1,1,-2)(-2,1,1)$	2
	$(-1,1,2)(2,1,-1)$	
5	$(-1,2,1)(1,2,-1)$	2
	$(-2,1,2)(2,1,-2)$	
6	$(-1,2,2)(2,2,-1)$	2
	$(1,2,-2)(-2,2,1)$	
7	$(-1,2,3)(3,1,-2)(-2,1,3)(3,2,-1)$	2
	$(1,2,-3)(-3,1,2)(2,1,-3)(-3,2,1)$	

Calculation for $\Delta = 17$: $\left[\sqrt{17}\right] = 4$, so possible values of b are 1,2,3,4.

b	$\Delta - b^2 = -ac$	$\sqrt{\Delta} + b$	$\sqrt{\Delta} - b$	forms
1	16	5.123	3.123	$(\pm 1,1,\mp 16)(\pm 2,1,\mp 8)(\pm 4,1,\mp 4)$
				$(\pm 16,1,\mp 1)(\pm 8,1,\mp 2)$
2	13	6.123	2.123	$(\pm 1,2,\mp 13)(\pm 13,2,\mp 1)$
3	8	7.123	1.123	$(\pm 1,3,\mp 8)(\pm 2,3,\mp 4)$
				$(\pm 8,3,\mp 1)(\pm 4,3,\mp 2)$
4	1	8.123	.123	$(\pm 1,4,\mp 1)$

6.65.9

The forms at the right satisfy $-ac = \Delta - b^2$. Using Proposition 3 of §64,

we may eliminate those which are unreduced, leaving

$$(\pm 4,1,\mp 4),(\pm 2,3,\mp 4)(\pm 4,3,\mp 2)(\pm 1,4,\mp 1)$$

There are various methods to determine the periods. We will do it

by using the first roots:

f $(4,1,-4)(-4,1,4)(2,3,-4)(-2,3,4)(4,3,-2)(-4,3\ 2)(1,4,-1)(-1,4,1)$

$$\alpha\qquad \frac{-1-\sqrt{17}}{4}\quad \frac{-1-\sqrt{17}}{-4}\quad \frac{-3-\sqrt{17}}{2}\quad \frac{-3-\sqrt{17}}{-2}\quad \frac{-3-\sqrt{17}}{4}\quad \frac{-3-\sqrt{17}}{-4}\quad \frac{-4-\sqrt{17}}{1}\quad \frac{-4-\sqrt{17}}{-1}\ .$$

We select a positive α and perform a continued fraction expansion on it

$$\frac{1+\sqrt{17}}{4} = (1,\frac{3+\sqrt{17}}{2}) = (1,3,\frac{3+\sqrt{17}}{4}) = (1,3,1,\frac{1+\sqrt{17}}{4})$$

$$= (1,3,1,1,3,1,\frac{1+\sqrt{17}}{4}).$$

Recalling that the first roots of adjacent forms have opposite signs, the

sequence of roots is

$$\frac{1+\sqrt{17}}{4}\ ,\ \frac{3+\sqrt{17}}{2}\ ,\ \frac{3+\sqrt{17}}{4}\ ,\ \frac{1+\sqrt{17}}{4}\ ,\ \frac{3+\sqrt{17}}{2}\ ,\ \frac{3+\sqrt{17}}{4}$$

and the period of forms is, going in the opposite order,

$$(4,3,-2),(-2,3,4),(4,1,-4),(-4,3,2),(2,3,-4),(-4,1,4).$$

There remain two reduced forms, and since a period contains an even

number of forms, these two form a period:

$$(-1,4,1),(1,4,-1).$$

There are thus two classes of forms with discriminant 17.

66. EXPANSION OF $\sqrt{\Delta}$ IN A CONTINUED FRACTION

Let Δ be a positive integer not the square of an integer. Then $\sqrt{\Delta}$ will be an irrational number and $\sqrt{\Delta} > 1$. We set $\lambda = [\sqrt{\Delta}]$, so $\lambda < \sqrt{\Delta} < \lambda + 1$ with $\lambda \geq 1$. We form the irrational numbers $\omega = \lambda + \sqrt{\Delta}$, $\eta = \lambda - \sqrt{\Delta}$. Thus $\omega > 1$ and $|\eta| < 1$. But ω and η are the first and second roots of the quadratic form

$$-x^2 + 2\lambda xy + (\Delta - \lambda^2)y^2 = (-1, \lambda, \Delta - \lambda^2)$$

which thus is reduced. Consequently, according to Proposition 3 of the preceeding section, we will have

$$\omega = \lambda + \sqrt{\Delta} = (2\lambda, \underbrace{q_1, q_2, \ldots, q_{2n-1}}_{\text{period}}, 2\lambda, q_1, \ldots) \tag{66.1}$$

and

$$\left|\frac{1}{\eta}\right| = \frac{1}{\sqrt{\Delta} - \lambda} = (\underbrace{q_{2n-1}, \ldots, q_2, q_1, 2\lambda}_{\text{period}}, q_{2n-1}, \ldots). \tag{66.2}$$

Then (66.1) implies

$$\sqrt{\Delta} = (\lambda, \underbrace{q_1, q_2, \ldots, q_{2n-1}, 2\lambda}_{\text{period}}, q_1, \ldots), \tag{66.3}$$

and, noticing that $\sqrt{\Delta} = \lambda + (\sqrt{\Delta} - \lambda) = \lambda + \dfrac{1}{1/(\sqrt{\Delta} - \lambda)}$, (66.2) implies

$$\sqrt{\Delta} = (\lambda, \underbrace{q_{2n-1}, \ldots, q_1, 2\lambda}_{\text{period}}, q_{2n-1}, \ldots). \tag{66.4}$$

Comparing (66.3) and (66.4) we see that

$$q_1 = q_{2n-1}, \quad q_2 = q_{2n-2}, \quad \cdots, \quad q_{2n-i} = q_i, \ldots, \qquad i = 1, 2, \ldots, n.$$

So we have arrived at the following proposition:

<u>PROPOSITION</u> The expansion of $\sqrt{\Delta}$ in a continued fraction has the form

$$\sqrt{\Delta} = (\lambda, \underbrace{q_1, q_2, \ldots, q_n, q_{n-1}, \ldots, q_2, q_1, 2\lambda}_{\text{period}}, \ldots).$$

6.66.2

EXAMPLES

$$\sqrt{6} = (2,2,4,2,4,\ldots)$$

$$\sqrt{7} = (2,1,1,1,4,1,1,\ldots)$$

$$\sqrt{8} = (2,1,4,1,4,1,\ldots)$$

$$\sqrt{10} = (3,6,6,6,\ldots)$$

$$\sqrt{23} = (4,1,3,1,8,1,\ldots)$$

$$\sqrt{29} = (5,2,1,1,2,10,\ldots)$$

REMARK The reader will perhaps be puzzled by the fact that the period in $\sqrt{29}$ has length 5.* Careful examination of the proof of the proposition reveals that it asserts a periodicity for $\sqrt{29}$ as follows

$$\sqrt{29} = (5,2,1,1,2,10,2,1,1,2,10).$$
$$|\text{———period———}|$$

---

*Additional material concerning continued fractions whose shortest period is odd will be found in the problems.

67. EQUIVALENCE OF A FORM WITH ITSELF AND THE SOLUTION OF THE EQUATION OF FERMAT FOR FORMS WITH POSITIVE DISCRIMINANT Δ

Let $f \sim f'$ and $fS_1 = f'$. If we know how to find all the unimodular substitutions T for which $fT = f$, then we will be able to determine all the unimodular substitutions which transform f into f'. Indeed, $fS = f'$ implies $fSS_1^{-1} = f'S_1^{-1} = f$ so $SS_1^{-1} = T$ and $S = TS_1$.

Let now $f = (a,b,c)$ be a quadratic form with positive discriminant not the square of an integer, and let $f\left(\begin{smallmatrix} \alpha & \beta \\ \gamma & \delta \end{smallmatrix}\right) = f$, so that also $f = f\left(\begin{smallmatrix} \delta & -\beta \\ -\gamma & \alpha \end{smallmatrix}\right)$. Then, as we saw in §57, there exist two integers t and u such that

$$t^2 - \Delta u^2 = \sigma^2 \tag{67.1}$$

where σ is the greatest common divisor of a, $2b$, and c and the following relations hold:

$$\alpha = \frac{t+bu}{\sigma} \qquad \beta = \frac{cu}{\sigma}$$

$$\gamma = \frac{-au}{\sigma} \qquad \delta = \frac{t-bu}{\sigma}. \tag{67.2}$$

From these we have

$$u = \frac{\beta\sigma}{c} \qquad t = \frac{\alpha+\delta}{2}\sigma.$$

Conversely, to each pair t,u satisfying (67.1) the formulas (67.2) determine a unimodular substitution $T = \left(\begin{smallmatrix} \alpha & \beta \\ \gamma & \delta \end{smallmatrix}\right)$ which transforms f into itself. To the pair $-t,-u$ then corresponds the unimodular substitution $\left(\begin{smallmatrix} -\alpha & -\beta \\ -\gamma & -\delta \end{smallmatrix}\right)$ which we will designate by $-T$.

Now let $f = (a,b,c)$ be a reduced quadratic form with positive discriminant not the square of an integer and with positive first root ω. Then we will have, according to §65,

$$\omega = (q_1, q_2, \ldots, q_{2n}, \omega)$$

6.67.2

where $q_1, q_2, \ldots, q_{2n}$ are positive integers, and we require $q_1, \ldots, q_{2n}$ to be the shortest sequence of even length with final denominator ω. Thus

$$\omega = \frac{Z_{2n}\omega + Z_{2n-1}}{N_{2n}\omega + N_{2n-1}} \, ,$$

so, if we set $\alpha_1 = Z_{2n}$, $\beta_1 = Z_{2n-1}$, $\gamma_1 = N_{2n}$, $\delta_1 = N_{2n-1}$ we will have

$$\omega = \frac{\alpha_1 \omega + \beta_1}{\gamma_1 \omega + \delta_1} \tag{67.3}$$

with $\alpha_1, \beta_1, \gamma_1, \delta_1$ positive integers and $\alpha_1\delta_1 - \beta_1\gamma_1 = (-1)^{2n} = 1$.

We will designate the unimodular substitution $\begin{pmatrix} \alpha_1 & \beta_1 \\ \gamma_1 & \delta_1 \end{pmatrix}$ by T_1; then we will have (Proposition 4, §56)

$$\omega = T_1\omega, \quad \text{so} \quad f\begin{pmatrix} \alpha_1 & \beta_1 \\ \gamma_1 & \delta_1 \end{pmatrix} = fT_1 = f.$$

Then we obviously have

$$\omega = T_1^m\omega \qquad fT_1^m = f \tag{67.4}$$

for every integer m.

To show the converse, we need the following

LEMMA Let $(q_1, q_2, \ldots)$ be a continued fraction and Z_n, N_n be as defined by (24.2) so that Z_n/N_n is the partial quotient. Let

$$Q_j = \begin{pmatrix} q_j & 1 \\ 1 & 0 \end{pmatrix}$$

and set $Z_0 = 1$ $N_0 = 0$.

Then

$$\begin{pmatrix} Z_n & Z_{n-1} \\ N_n & N_{n-1} \end{pmatrix} = Q_1 Q_2 \ldots Q_n \qquad \text{for } n \geqslant 1.$$

Proof This is clear for n = 1. If it is true for k then

$$Q_1 Q_2 \ldots Q_k Q_{k+1} = \begin{pmatrix} Z_k & Z_{k-1} \\ N_k & N_{k-1} \end{pmatrix}\begin{pmatrix} q_{k+1} & 1 \\ 1 & 0 \end{pmatrix}$$

$$= \begin{pmatrix} q_{k+1}Z_k + Z_{k-1} & Z_k \\ q_{k+1}N_k + N_{k-1} & N_k \end{pmatrix} = \begin{pmatrix} Z_{k+1} & Z_k \\ N_{k+1} & N_k \end{pmatrix},$$

so it is true for k+1. Thus the lemma holds for all k.

6.67.3

<u>COROLLARY</u> Let a continued fraction be periodic with period 2n.

Then $\begin{pmatrix} Z_{2rn} & Z_{2rn-1} \\ N_{2rn} & N_{2rn-1} \end{pmatrix} = \begin{pmatrix} Z_{2n} & Z_{2n-1} \\ N_{2n} & N_{2n-1} \end{pmatrix}^r .$

<u>Proof</u> By the lemma

$$\begin{pmatrix} Z_{2rn} & Z_{2rn-1} \\ N_{2rn} & N_{2rn-1} \end{pmatrix} = Q_1 \ldots Q_{2n} Q_1 \ldots Q_{2n} \ldots Q_1 \ldots Q_{2n}$$
$$|\text{------}r \text{ periods------}|$$
$$= \begin{pmatrix} Z_{2n} & Z_{2n-1} \\ N_{2n} & N_{2n-1} \end{pmatrix} \begin{pmatrix} Z_{2n} & Z_{2n-1} \\ N_{2n} & N_{2n-1} \end{pmatrix} \ldots \begin{pmatrix} Z_{2n} & Z_{2n-1} \\ N_{2n} & N_{2n-1} \end{pmatrix}$$
$$= \begin{pmatrix} Z_{2n} & Z_{2n-1} \\ N_{2n} & N_{2n-1} \end{pmatrix}^r .$$

We will now show the converse; that every unimodular substitution $T = \begin{pmatrix} \alpha & \beta \\ \gamma & \delta \end{pmatrix}$ which transforms f into itself is equal to $\pm T_1^m$ for some integer m. The hypothesis $fT = f\begin{pmatrix} \alpha & \beta \\ \gamma & \delta \end{pmatrix} = f$ implies that

$$\omega = \frac{\alpha\omega+\beta}{\gamma\omega+\delta} . \tag{67.5}$$

According to the theorem of Lagrange (§63) we may determine k and h such that

$$\omega = (q_1,\ldots,q_k,\omega_k) \quad \text{and} \quad \omega = (r_1,\ldots,r_h,\omega'_h), \tag{67.6}$$

where $\omega_k = \omega'_h$ and (67.5) results from the elimination of $\omega_k = \omega'_h$ from (67.6). By the remark following the Theorem of Lagrange, we may take k to be any integer exceeding a certain natural number c and since $\alpha\delta - \beta\gamma = 1$ the corresponding h will satisfy $k \equiv h \pmod 2$. We select $k = 2rn > c$. Then

$$\omega = (\overset{|1\text{st period}|}{q_1}, \ldots, q_{2n}, \overset{|2\text{nd period}|}{q_1}, \ldots, q_{2n}, \ldots, \overset{|r\text{th period}|}{q_1}, \ldots, q_{2n}, \omega_{2rn}),$$

so $\omega_{2rn} = \omega$. We claim $h = 2sn$ for some s, so that

$$\frac{\alpha\omega+\beta}{\gamma\omega+\delta} = \omega = (\overset{|1\text{st period}|}{q_1}, \ldots, q_{2n}, \ldots, \overset{|s\text{th period}|}{q_1}, \ldots, q_{2n}, \omega_{2sn})$$

where $\omega_{2sn} = \omega$.

6.67.4

Indeed, since $\omega_h^* = \omega_k = \omega_{2rn} = \omega$, ω_h^* can occur at positions 2sn or, possibly, at positions 2sn+n. In the latter case, however, n must be odd, since otherwise n and not 2n would be the length of the shortest period of ω. But if n is odd, $2sn+n \not\equiv 2rn \pmod 2$ as required by the theorem of Lagrange (Proposition 3 of §63). Hence h = 2sn.

We now have

$$\omega = \begin{pmatrix} Z_{2rn} & Z_{2rn-1} \\ N_{2rn} & N_{2rn-1} \end{pmatrix} \omega_{2rn},$$

$$\frac{\alpha\omega+\beta}{\gamma\omega+\delta} = \begin{pmatrix} Z_{2sn} & Z_{2sn-1} \\ N_{2sn} & N_{2sn-1} \end{pmatrix} \omega_{2sn}$$

and, using the Corollary to the theorem of Lagrange, we have

$$\pm\begin{pmatrix} \alpha & \beta \\ \gamma & \delta \end{pmatrix} = \begin{pmatrix} Z_{2sn} & Z_{2sn-1} \\ N_{2sn} & N_{2sn-1} \end{pmatrix} \begin{pmatrix} Z_{2rn} & Z_{2rn-1} \\ N_{2rn} & N_{2rn-1} \end{pmatrix}^{-1}. \tag{67.7}$$

However, by the corollary to the last lemma this becomes

$$\pm\begin{pmatrix} \alpha & \beta \\ \gamma & \delta \end{pmatrix} = T_1^{\,s}(T_1^{\,r})^{-1} = T_1^{\,s-r}.$$

Thus we have the:

PROPOSITION 1 All the unimodular transformations T which transform f into itself have the form $T = \pm T_1^{\,m}$ for some integer m.

We have now completely solved the second problem of §53, to which the second basic problem of §46 had been reduced, (for positive quadratic forms with discriminant not the square of an integer).

Using equations (67.2), it is possible to calculate all the T which transform f into itself in another way, which is closely connected with the equation of Fermat, to which we now turn our attention.

6.67.5

Since the integer solutions of the equation $t^2 - \Delta u^2 = \sigma^2$ correspond in a one to one manner with the unimodular substitutions T which transform f into itself, that is, with $\pm T_1{}^m$ (m an integer), the above results furnish a means for the complete solution of the equation of Fermat.

Let us call (t_1, u_1) the solution which corresponds to the substitution $T_1 = \begin{pmatrix} \alpha_1 & \beta_1 \\ \gamma_1 & \delta_1 \end{pmatrix}$; then we have $u_1 = \dfrac{\beta_1 \sigma}{c} > 0$ since $\beta_1 > 0$ and $c > 0$; (recall that (a,b,c) had a positive first root so $c > 0$). In addition, $t_1 = \dfrac{\alpha_1 + \delta_1}{2} \sigma > 0$.

Consider now two substitutions T and T' such that
$$fT = f \quad \text{and} \quad fT' = f.$$
Let (t,u) and (t',u') be the solutions of the equation of Fermat corresponding to T and T'. We set
$$TT' = T''.$$
We pose the question, how does the pair (t'',u'') corresponding to T'' relate to the pairs (t,u) and (t',u') of T and T'?

We have
$$\begin{pmatrix} \alpha'' & \beta'' \\ \gamma'' & \delta'' \end{pmatrix} = T'' = TT' = \begin{pmatrix} \dfrac{t+bu}{\sigma} & \dfrac{cu}{\sigma} \\ \dfrac{-au}{\sigma} & \dfrac{t-bu}{\sigma} \end{pmatrix} \begin{pmatrix} \dfrac{t'+bu'}{\sigma} & \dfrac{cu'}{\sigma} \\ \dfrac{-au'}{\sigma} & \dfrac{t'-bu'}{\sigma} \end{pmatrix} \tag{67.8}$$
with
$$\beta'' = \frac{t+bu}{\sigma}\,\frac{cu'}{\sigma} + \frac{cu}{\sigma}\,\frac{t'-bu'}{\sigma} = \frac{c}{\sigma^2}\,(tu'+ut')$$
so
$$u'' = \frac{\beta'' \sigma}{c} = \frac{1}{\sigma}\,(tu'+t'u).$$
Moreover,

6.67.6

$$\alpha'' = \frac{t+bu}{\sigma}\,\frac{t'+bu'}{\sigma} + \frac{cu}{\sigma}\left(\frac{-au'}{\sigma}\right) = \frac{1}{\sigma^2}\left[(t+bu)(t'+bu')-acuu'\right]$$

$$\delta'' = \left(\frac{-au}{\sigma}\right)\left(\frac{cu'}{\sigma}\right) + \frac{t-bu}{\sigma}\,\frac{t'-bu'}{\sigma} = \frac{1}{\sigma^2}\left[(t-bu)(t'-bu')-acuu'\right]$$

so

$$t'' = \frac{\alpha''+\delta''}{2}\,\sigma = \frac{1}{2\sigma}\left[2tt'+2b^2uu'-2acuu'\right] = \frac{1}{\sigma}\left[tt'+\Delta uu'\right].$$

Consequently

$$u'' = \frac{ut'+tu'}{\sigma} \qquad t'' = \frac{tt'+\Delta uu'}{\sigma} \tag{67.9}$$

and thus

$$\frac{t''+u''\sqrt{\Delta}}{\sigma} = \frac{tt'+\Delta uu'+(ut'+tu')\sqrt{\Delta}}{\sigma^2}$$

$$= \frac{(t+u\sqrt{\Delta})(t'+u'\sqrt{\Delta})}{\sigma^2}.$$

So, finally, we have found

$$\frac{t''+u''\sqrt{\Delta}}{\sigma} = \frac{t+u\sqrt{\Delta}}{\sigma}\,\frac{t'+u'\sqrt{\Delta}}{\sigma}. \tag{67.10}$$

From this relation we may recover the relations (67.9) by splitting the

product on the right side into rational and irrational parts.

Specializing to the case $T' = T^{-1}$, we have

$T'' = \left(\begin{smallmatrix} 1 & 0 \\ 0 & 1 \end{smallmatrix}\right)$, so $u'' = 0$ and $t'' = \sigma$. Consequently (67.10) becomes

$$1 = \frac{t+u\sqrt{\Delta}}{\sigma}\cdot\frac{t'+u'\sqrt{\Delta}}{\sigma}.$$

From this relation we immediately derive

$$\frac{t-u\sqrt{\Delta}}{\sigma} = \frac{t-u\sqrt{\Delta}}{\sigma}\,\frac{t+u\sqrt{\Delta}}{\sigma}\,\frac{t'+u'\sqrt{\Delta}}{\sigma}$$

$$= \frac{t^2-\Delta u^2}{\sigma^2}\,\frac{t'+u'\sqrt{\Delta}}{\sigma}$$

$$= 1\cdot\frac{t'+u'\sqrt{\Delta}}{\sigma},$$

and by separation into rational and irrational parts we find

$$t' = t, \quad u' = -u.$$

Thus if T corresponds to the pair (t,u), then T^{-1} corresponds to the pair

$(t,-u)$.

Now let m be a natural number. We take

$$\underbrace{T_1 T_1 \ldots T_1}_{m \text{ times}} = T_1^{\,m}$$

where T_1 is the substitution we defined at the beginning of this section. If we designate by (t_1, u_1) the pair corresponding to T_1 and by (t_m, u_m) the pair corresponding to $T = T_1^{\,m}$, then with repeated use of (67.10) we will get

$$\frac{t_m + u_m \sqrt{\Delta}}{\sigma} = \left(\frac{t_1 + u_1 \sqrt{\Delta}}{\sigma}\right)^m. \tag{67.11}$$

Let now m be a negative integer and $m = -n$ with $n > 0$. We have

$$T_1^{\,m} = T_1^{\,-n} = (T_1^{\,n})^{-1} \text{ so } T_1^{\,m} T_1^{\,n} = \begin{pmatrix} 1 & 0 \\ 0 & 1 \end{pmatrix}. \tag{67.12}$$

However, by (67.11)

$$\frac{t_n + u_n \sqrt{\Delta}}{\sigma} = \left(\frac{t_1 + u_1 \sqrt{\Delta}}{\sigma}\right)^n.$$

Let us set

$$t_m = t_n \qquad u_m = -u_n.$$

Then, by (67.12) and the above rule for inverses,

$$\frac{t_m - u_m \sqrt{\Delta}}{\sigma} = \frac{t_n + u_n \sqrt{\Delta}}{\sigma} = \left(\frac{t_1 + u_1 \sqrt{\Delta}}{\sigma}\right)^n = \left(\frac{t_1 + u_1 \sqrt{\Delta}}{\sigma}\right)^{-m};$$

so

$$\frac{t_m + u_m \sqrt{\Delta}}{\sigma} = \left(\frac{t_m - u_m \sqrt{\Delta}}{\sigma}\right)^{-1} = \left(\frac{t_1 + u_1 \sqrt{\Delta}}{\sigma}\right)^m$$

and forumla (67.11) is valid for $m = 0, \pm 1, \pm 2, \ldots$.

We have now

PROPOSITION 2   The most general solution of the equation of Fermat $t^2 - \Delta u^2 = \sigma^2$ with $\sigma = 1, 2$ and positive Δ not the square of an integer is given by the formula

$$\frac{t + u \sqrt{\Delta}}{\sigma} = \pm \left(\frac{t_1 + u_1 \sqrt{\Delta}}{\sigma}\right)^m \quad \text{with m an integer.}$$

6.67.8

REMARK Both t and u are positive when $m > 0$ and the positive sign is chosen, and indeed $t > t_1$ and $u > u_1$ when $m > 1$.

REMARK The unimodular T for the forms of discriminant $D = -\Delta$ are determined from the t,u by (67.2).

EXAMPLES I. Solve the equation $t^2 - 6u^2 = \sigma^2 = 1$.

We find a reduced quadratic form with discriminant 6, a positive first root ω, and divisor $\sigma = 1$; for example, $(-1,2,2)$.* We expand the first root $\omega = 2+\sqrt{6}$ in a continued fraction. We have

$$\omega = 2+\sqrt{6} = 4+(\sqrt{6}-2) = 4+\cfrac{1}{(\sqrt{6}+2)/2}$$

$$\frac{\sqrt{6}+2}{2} = 2+\frac{\sqrt{6}-2}{2} = 2+\cfrac{1}{\sqrt{6}+2} \,,$$

so

$$\omega = 2+\sqrt{6} = (4,2,2+\sqrt{6}) = (4,2,\omega).$$

We have

$$\begin{pmatrix} Z_2 & Z_1 \\ N_2 & N_1 \end{pmatrix} = \begin{pmatrix} 9 & 4 \\ 2 & 1 \end{pmatrix} \quad \text{and} \quad \begin{pmatrix} 9 & 4 \\ 2 & 1 \end{pmatrix}(2+\sqrt{6}) = 2+\sqrt{6}.$$

Thus

$$T_1 = \begin{pmatrix} \alpha_1 & \beta_1 \\ \gamma_1 & \delta_1 \end{pmatrix} = \begin{pmatrix} 9 & 4 \\ 2 & 1 \end{pmatrix}$$

and

$$t_1 = \frac{9+1}{2} = 5 \quad \text{and} \quad u_1 = \frac{4 \cdot 1}{2} = 2$$

is the solution of $t^2 - 6u^2 = 1$ which corresponds to T_1. All the solutions of $t^2 - 6u^2 = 1$ are then given by the formula

$$t+u\sqrt{6} = \pm(5+2\sqrt{6})^m \qquad \text{where } m \in \mathbb{Z}.$$

*This can always be done by reducing the form $(1,0,-\Delta)$.

6.67.9

    II. Solve the equation $t^2-17u^2 = \sigma^2 = 4$.

We take a reduced form with $\Delta = 17$, $\sigma = 2$, and positive first root;

for instance $(-4,3,2)$. We expand its first root $\omega = \dfrac{3+\sqrt{17}}{4}$ in a continued

fraction:

$$\frac{3+\sqrt{17}}{4} = 1+\frac{\sqrt{17}-1}{4} = 1+\frac{1}{(\sqrt{17}+1)/4} \; ,$$

$$\frac{\sqrt{17}+1}{4} = 1+\frac{\sqrt{17}-3}{4} = 1+\frac{1}{(\sqrt{17}+3)/2} \; ,$$

$$\frac{\sqrt{17}+3}{2} = 3+\frac{\sqrt{17}-3}{2} = 3+\frac{1}{(\sqrt{17}+3)/4} \; ,$$

so

$$\omega = \frac{3+\sqrt{17}}{4} = (1,1,3,\frac{\sqrt{17}+3}{4}) \; .$$

Since the period has odd length, we must go another period to get a

unimodular substitution (the determinant must be +1). So

$$\omega = \frac{3+\sqrt{17}}{4} = (1,1,3,1,1,3,\frac{\sqrt{17}+3}{4}).$$

We calculate $T_1 = \begin{pmatrix} Z_6 & Z_5 \\ N_6 & N_5 \end{pmatrix}$,

$Z_1 = 1$, $Z_2 = 2$, $Z_3 = 7$, $Z_4 = 9$, $Z_5 = 16$, $Z_6 = 57$

$N_1 = 1$, $N_2 = 1$, $N_3 = 4$, $N_4 = 5$, $N_6 = 9$, $N_7 = 32$

so

$$T_1 = \begin{pmatrix} 57 & 16 \\ 32 & 9 \end{pmatrix}.$$

The pair t_1, u_1 corresponding to T_1 is

$$t_1 = \frac{57+9}{2} \cdot 2 = 66 \qquad u_1 = \frac{16 \cdot 2}{2} = 16.$$

Thus $(66,16)$ is a solution of $t^2-17u^2 = 4$, and all solutions of this

equation of Fermat are given by the formula

$$t+u\sqrt{17} = \pm(66+16\sqrt{17})^m \qquad \text{with } m \in \mathbf{Z}.$$

PROBLEMS FOR CHAPTER 6

1. Making use of the formula $af(x,y) = (ax+by)^2 - \Delta y^2$ (§47), find all
   solutions of $2x^2+6xy+5y^2 = 1$.

2. Solve $29x^2-82xy+58y^2 = 1$ by subjecting this quadratic form to the linear
   transformation $x = 3x'+7y'$, $y = 2x'+5y'$, using the matrix technique in
   §49.

3. Find the auxiliary forms described in §53 needed to solve the following
   problems

   a) $2x^2+6xy+7y^2 = 11$.

   b) $2x^2+6xy+7y^2 = 27$.

   c) $2x^2+6xy+6y^2 = 98$.

   d) $2x^2+10xy+4y^2 = 94$.

   e) $x^2+10xy+8y^2 = 53$.

   f) $x^2+10xy+8y^2 = 74$.

4. Find all reduced forms of discriminant $\Delta = -14$ ($D = 14$).

5. Reduce the forms $2x^2+6xy+7y^2$ and $2x^2+6xy+6y^2$ using the method of
   proposition 1 of §54. Find the matrices which transform these forms into
   their reduced equivalents. What is the smallest positive integer
   representable by each of these forms?

   We will in some of the following problems make use of the concept of
   improper equivalence, which we symbolize by $f \sim f'$. We define $f \sim f'$ if and
   only if there is an S with det $S = -1$ for which $fS = S'$. It is possible for
   $f' = S_1 f$ with det $S_1 = 1$ and $f' = S_2 f$ with det $S_2 = -1$, so that S is both
   (properly) equivalent and improperly equivalent to S'.

6.P.2

6. Let ω and η be the first and second roots respectively of $f = (a,b,c)$, and ω_1, η_1 the first and second roots respectively of $f_1 = (c,b,a)$. Assume $ac \neq 0$. that $\omega_1 = \dfrac{1}{\eta}$, $\eta_1 = \dfrac{1}{\omega}$.

7. Show that $f = (a,b,c) \simeq (c,b,a) = f'$ using $S_0 = \begin{pmatrix} 0 & 1 \\ 1 & 0 \end{pmatrix}$. Show that $\omega = S_0\eta'$, where ω is the <u>first</u> root of f and η' the <u>second</u> root of f'. Also show $\eta = S_0\omega'$.

8. Show that if $f \simeq f'$ using the transformation S then $S\eta' = \omega$, $S\omega' = \eta$. Hint: combine S with S_0 of problem 7.

9. Let f and f' have the same discriminant, $aa'c' \neq 0$, and $\omega = S\eta'$, $\eta = S\omega'$, where det $S = -1$. Show $f \simeq f'$. . Hint; consider SS_0 and use Prop 4 of §56.

10. Show that $f = (a,b,a)$ is improperly equivalent to itself.

11. Let $(a,b,c)\begin{pmatrix} \alpha & \beta \\ \gamma & \delta \end{pmatrix} = (a',b',c')$, where det $\begin{pmatrix} \alpha & \beta \\ \gamma & \delta \end{pmatrix} = +1$. Show, using problem 6, that

$(c',b',a')\begin{pmatrix} \alpha & -\gamma \\ -\beta & \delta \end{pmatrix} = (c,b,a)$.

12. Find the reduced forms corresponding to the original and the auxiliary forms of problems 3a, b, c, and thus determine solvability of the problems.

13. Use the methods of §59 to find all transformations of the original forms of 3a, b, c into themselves, and thus determine all solutions of those of 3a, b, c which are solvable.

The following problems 12-15 concern various aspects of the representation of an integer m by the form x^2+y^2. Notation is as in §61.

14. Show that $A(4m) = A(m)$ by showing that every solution $m = x^2+y^2$ generates a solution $4m = x'^2+y'^2$ and conversely.

6.P.3

15. Show that if m is odd then $A(2m) = A(m)$, by exhibiting a one to one
    correspondence between the solutions for m and those for $2m$. Hint: Show
    first that if $2m = x^2+y^2$ then x and y are both odd. Set $x' = \frac{x-y}{2}$ and
    show that x' is part of a solution $m = x'^2+y'^2$. Then set up the one to
    one correspondence.

16. Show, using problems 14 and 15 that formula (61.6) modified by the
    condition that δ be an <u>odd</u> divisor remains valid when m is an even
    integer.

17. Let $m_1 = x_1^2 + y_1^2$ and $m_2 = x_2^2 + y_2^2$. Show that a representation of $m_1 m_2$
    may be found by multiplying the complex numbers x_1+iy_1 and x_2+iy_2.
    Hint: Consider the norm $N(x+iy) = |x+iy|^2 = x^2+y^2 = (x+iy)(x-iy)$.

The following problems 18-19 concern representations of primes $p \equiv 1 \pmod 3$.
The reader will find it useful to recall the form $\left(\frac{q^*}{p}\right) = \left(\frac{p}{q}\right)$ of the law of
quadratic reciprocity (Problem 20 of Chapter 5).

18. a) Show that a prime $p \neq 3$ can be represented in the form $p = a^2-ab+b^2$
        if and only if $p \equiv 1 \pmod 3$. Hint: We must consider
        $2p = 2a^2 - 2ab+2b^2$. This is a form of the second kind.

    b) Find all the unimodular transformations S which transform $f(x,y) =$
        $2x^2-2xy+2y^2 = (2,-1,2)$ into itself, (see §59).

    c) Show that there are exactly twelve distinct representations of p in
        the form a^2-ab+b^2. By removing the "trivial" distinctions
        $(a,b) \to (-a,-b)$ and $(a,b) \to (b,a)$, show that there are three
        essentially different representations, and use the matrices S from b)
        to find formulas for the two others in terms of a given one (a,b).

6.P.4

d) Show that, for a representation $p = a^2-ab+b^2$, $p \neq 3$, 3 divides exactly one of a, b, a-b. Conclude that there are four distinct representations (a,b) for which $3|b$.

19. Show, using problem 18d that for $p \neq 3$ there exists a representation of $4p = x^2+27y^2$ if and only if $p \equiv 1 \pmod 3$. Show that this representation is essentially unique (that is, there are four distinct representations). Hint: Rewrite $4p = 4a^2-4ab+4b^2$ as A^2+3b^2 and use 18d. For uniqueness, show that each representation of $4p = x^2+27y^2$ generates a unique representation of the type found in 18d.

20. Show that if a form (a,b,c) with positive discriminant Δ not the square of an integer is reduced, then (c,b,a) is also reduced. Hint: Consider the definition of reduced form in terms of its roots and use problem 6.

21. Show that a form (a,b,c) with positive discriminant Δ not the square of an integer is reduced if and only if $\sqrt{\Delta}+b > |c| > \sqrt{\Delta}-b > 0$. Hint: Use problem 20.

22. Show that (a',b',a'') is right adjacent to (a,b,a') if and only if (a'',b',a') is left adjacent to (a',b,a). Hint: Use problem 11.

23. Prove Proposition 2 of §65 from Proposition 1 of §65 and problems 20 and 22.

The following problems 24-29 treat the case of a quadratic irrational ω whose shortest period is odd. Let $\omega = (\overline{q_1 q_2 \cdots q_{2\ell+1}})$.

24. a) Show that a reduced ω with odd period is improperly equivalent to itself.

b) Find an improper equivalence of $(3+\sqrt{17})/4$ with itself.

6.P.5

25. a) Let ω be reduced with odd period, and ω be the first root

   of (a,b,c). Show that (a,b,c) is improperly equivalent to

   (-a,-b,-c). Hint: Let $\omega = S\omega$ be the improper equivalence given by

   problem 24, and $(a,b,c)S = (a',b',c')$. Consider the relation

   $\omega = S\omega$ in conjuction with problem 8.

   b) Find an improper equivalence of (-4,3,2) and (4,-3,-2).

26. Let (a,b,c) be a form with positive discriminant not the square of an

   integer whose first root ω has a odd period; $\omega = (a_1,\ldots,a_r,\overline{q_1,\ldots,q_{2\ell+1}})$.

   Show (a,b,c) is improperly equivalent to (-a,-b,-c).

27. a) Let $(a,b,c)S = (-a,-b,-c)$ be an improper equivalence. Show that the

   t and u defined in §57 are integers and satisfy $t^2 - \Delta u^2 = -\sigma^2$.

   b) Show that if there exist t and u satisfying $t^2 - \Delta u^2 = -\sigma^2$, where σ is

   the divisor of (a,b,c) then there is an improper equivalence $(a,b,c)S$

   $= (-a,-b,-c)$. Hint: Refer to the formulas in §57.

28. Show that if $t^2 - \Delta u^2 = -\sigma^2$ (where Δ is positive and not the square of an

   integer and $\sigma = 1$ or 2) is solvable, then any quadratic form (a,b,c) with

   this Δ and σ will have a first root with odd period. Hint: Use 27b, and

   construct an improper equivalence of ω with itself. Then use Lagrange's

   Theorem to relate the improper equivalence to the continued fraction

   expansion of ω.

29. Let (a,b,c) be a reduced form with positive first root ω not the square

   of an integer, where ω has an odd period. Let n be the minimal (odd)

   period and set $S_1 = \begin{pmatrix} Z_n & Z_{n-1} \\ N_n & N_{n-1} \end{pmatrix}$. Show that

   a) If $(a,b,c)S = \pm(a,b,c)$ then there is an integer j for which $S = \pm S_1^{\,j}$.

6.P.6

   b) If $(a,b,c)S = +(a,b,c)$ then the j in a) is even.

   c) The T_1 of §67 satisfies $T_1 = S_1{}^2$.

The following problems 30-32 contain additional facts about the equation of Fermat (and of Pell and Lagrange). Let the continued fraction expansion of $\sqrt{\Delta} = (q_0, \overline{q_1, \ldots, q_{2n}})$. where $q_1, \ldots,\ q_{2n}$ is the minimal period of <u>even</u> length and $\Delta > 1$. We also consider, in the case of $\sqrt{\Delta}$ possessing an odd period, the equation $t^2 - \Delta u^2 = -1$.

30. a) Let $q_0 = [\sqrt{\Delta}]$ and $Q_0 = \begin{pmatrix} q_0 & 1 \\ 1 & 0 \end{pmatrix}$. Find $f_1 = (1, 0, -\Delta)Q_0$ and show that it

      is reduced and has a positive first root ω_1.

   b) Let $\omega = (\overline{q_1, \ldots, q_{2n}})$, $Q_i = \begin{pmatrix} q_i & 1 \\ 1, & 0 \end{pmatrix}$. Then (see § 67), all

      transformations taking f_1 into itself are $\pm T_1{}^j$, where $T_1 =$

      $Q_1 \ldots Q_{2n}$. Show that all transformations taking $(1, 0, -\Delta)$ into itself

      are $\pm Q_0 T_1{}^j Q_0{}^{-1} = \pm U_j$, where j may be negative or 0.

   c) Each $T_1{}^j$ is correlated with a pair (t_j, u_j). What is U_j in terms of

      t_n, u_n?

   d) Let t, u be a solution of $t^2 - \Delta u^2 = 1$. Show that $S = \begin{pmatrix} t & \Delta u \\ u & t \end{pmatrix}$

      satisfies $(1, 0, -\Delta)S = (1, 0, -\Delta)$ and thus $S = \pm U_j$ for some j.

31. We wish to find formulas for t_j, u_j in terms of the convergents Z_0/N_0, Z_1/N_1, $Z_2/N_2, \ldots$ of $\sqrt{\Delta} = (q_0, \overline{q_1, \ldots,\ q_{2n}})$. Note that, with the notation of §67, $\begin{pmatrix} Z_j & Z_{j-1} \\ N_j & N_{j-1} \end{pmatrix} = Q_0\, Q_1 \ldots Q_j$ and $T_1 = Q_1 \ldots Q_{2n}$ (since T_1 is constructed from $q_1, \ldots, q_{2n}$ which are the quotients for the reduced ω_1).

   a) Show that $U_j = \begin{pmatrix} Z_{2nj-1} & Z_{2nj-2} \\ N_{2nj-1} & N_{2nj-2} \end{pmatrix} Q_{2nj}\, Q_0{}^{-1}$

      and, by calculating this value, that $t_j = Z_{2nj-1}$, $u_j = N_{2nj-1}$.

6.P.7

b) Show that any solution (t,u) of $t^2 - \Delta^2 = 1$ with $t > u > 0$ satisfies
$t = t_j = Z_{2nj-1}$, $\quad U = U_j = N_{2nj-1}$ for some $j > 0$.

c) Find, using this method, two solutions of $t^2 - 17u^2 = 1$ and two solutions of $t^2 - 7u^2 = 1$, where t and u are positive integers.

32. This problem treats the equation $r^2 - \Delta s^2 = -1$. We assume that $\Delta = (q_0, \overline{q_1, \ldots, q_n})$ where n is odd, and set $r_j = Z_{nj-1}$ and $s_j = N_{nj-1}$ where Z_0/N_0, Z_1/N_1, $Z_2/N_2 \ldots$ are the partial quotients of $\sqrt{\Delta}$.

a) Let S_1 be the improper equivalence generated by $\omega_1 = (q_1, \ldots, q_n, \omega_1)$. Show $S_1^2 = T_1$ and $(1,0, -\Delta)0_0 S_1 0_0^{-1} = (-1, 0, + \Delta)$.

b) Show that $V_1 = 0_0 S_1 0_0^{-1} = \begin{pmatrix} Z_{n-1}, & q_0 Z_{n-1} + Z_{n-2} \\ N_{n-1}, & q_0 N_{n-1} + N_{n-2} \end{pmatrix} = \begin{pmatrix} \alpha & \beta \\ \gamma & \delta \end{pmatrix}$. Use (48.2) to show $Z_{n-1}^2 - \Delta N_{n-1}^2 = -1$.

c) Show that if $\det \begin{pmatrix} \alpha & \beta \\ \gamma & \delta \end{pmatrix} = 1$ and $(1,0,-\Delta) \begin{pmatrix} \alpha & \beta \\ \gamma & \delta \end{pmatrix} = (1,0,-\Delta)$ or if $\det \begin{pmatrix} \alpha & \beta \\ \gamma & \delta \end{pmatrix} = -1$ and $(1,0,-\Delta) \begin{pmatrix} \alpha & \beta \\ \gamma & \delta \end{pmatrix} = (-1, 0, \Delta)$ then $\beta = \Delta \cdot \gamma$ and $\delta = \alpha$.

d) Show that $V_1^j = \begin{pmatrix} Z_{nj-1} & q_0 Z_{nj-1} + Z_{nj-2} \\ N_{nj-1} & q_0 N_{nj-1} + N_{nj-2} \end{pmatrix} = \begin{pmatrix} Z_{nj-1}, & Z_{nj} - q_0 Z_{nj-1} \\ N_{nj-1}, & N_{nj} - q_0 N_{nj-1} \end{pmatrix}$

$= \begin{pmatrix} Z_{nj-1}, & \Delta N_{nj-1} \\ N_{nj-1}, & Z_{nj-1} \end{pmatrix}$,

and that
$$(1,0,-\Delta) V^j = \begin{cases} (1,0,-\Delta) & \text{for } j \text{ even} \\ (-1,0,\Delta) & \text{for } j \text{ odd}. \end{cases}$$

6.P.8

e) Show, using equation 48.2, that $Z_{nj-1}^2 - \Delta N_{nj-1}^2 = (-1)^j$.

f) Show that $Z_{nj-2}^2 - \Delta N_{nj-2}^2 = Z_{nj}^2 - \Delta N_{nj}^2 = (-1)^j (q_0^2 - \Delta)$.

g) Show that any solution (r,s) of $r^2 - \Delta s^2 = (-1)^k$ with $r,s > 0$ satisfies $r = Z_{nj-1}$, $s = N_{nj-1}$ for some j. Hint: we know this if k is even. If k is odd, show $(r_1 + s_1 \sqrt{\Delta}) = (r + s \sqrt{\Delta})^2$ satisfies $r_1^2 - \Delta s_1^2 = +1$ and thus $r_1 = Z_{2nj-1}$, $s_1 = N_{2nj-1}$. Consider $r_2 = Z_{nj-1}$, $s_2 = N_{nj-1}$, and show (using $V_1{}^j$) that $(r_2 + s_2 \sqrt{\Delta})^2 = r_1 + s_1 \sqrt{\Delta}$ from which it is easy to show that $r = r_2$, $s = s_2$.

h) Find, using these methods, two solutions of $x^2 - 17y^2 = -1$ and one solution of $x^2 - 13y^2 = -1$.

The following problems 33-40 treat some interesting properties of the continued fraction expansions of $\sqrt{\frac{A}{B}}$ where A and B are relatively prime and $A > B \geqslant 1$ and A/B is not a square. We set

$\sqrt{\frac{A}{B}} = (q_0, \overline{q_1, \ldots, q_m})$ where $q_0 = [\sqrt{\frac{A}{B}}]$ and $q_1, \ldots, q_m$ is the shortest period, odd or even. We set (see problem 33)

$\sqrt{\frac{A}{B}} = (q_0, q_1, \ldots, q_j, \omega_j)$, so that $\omega_n = \omega_0$, and will show that $q_m = 2q_0$ and ω_j is the first root of a reduced form for $j > 0$. We also set

$\omega_j = (r_j + \sqrt{\Delta})/s_j$ and $\Delta = AB$.

33. Show that the Proposition of §66 remains valid if $\sqrt{\Delta}$ is replaced by $\sqrt{\frac{A}{B}}$, where $(A,B) = 1$ and $A > B > 0$. Set $\Delta = AB$ so $\sqrt{\frac{A}{B}} = \frac{\sqrt{\Delta}}{B}$ and proceed as in §66.

6.P.9

34. Using the fact that ω_j is the first root of a reduced form and $\omega_j > 0$ show that $0 < r_j < \sqrt{\Delta}$ and $0 < s_j < 2\sqrt{\Delta}$ for $j > 0$.

35. Expand $\sqrt{\dfrac{13}{7}}$, determining q_j, r_j, s_j. Here is a convenient algorithm:

$$\sqrt{\frac{13}{7}} = \frac{\sqrt{91}}{7} = 1 + \frac{\sqrt{91}-7}{7} = 1 + \frac{42}{7(\sqrt{91}+7)}$$

$$\omega_0 = \frac{\sqrt{91}+7}{6} = 2 + \frac{\sqrt{91}-5}{6} = 2 + \frac{66}{6(\sqrt{91}+5)}$$

$$\omega_1 = \frac{\sqrt{91}+5}{11} = 1 + \frac{\sqrt{91}-6}{11} = 1 + \frac{55}{11(\sqrt{91}+6)} \quad \text{etc.}$$

36 a) Show that, as a consequence of the continued fraction algorithm the following relations hold for $j > 0$

$$r_{j+1} = q_{j-1}\, s_j - r_j$$

$$s_{j+1} = \frac{\Delta - r_j^{\,2}}{s_j} + 2r_j q_{j+1} - s_j q_{j+1}^{\,2}$$

and thus

$$r_j + r_{j+1} = q_{j+1}\, s_j$$
$$s_j\, s_{j+1} = \Delta - r_{j+1}^{\,2}$$
$$s_{j-1}\, s_j + q_{j+1}\, r_j s_j = \Delta + r_j r_{j+1}$$

Show that all but the last hold for $j = -1$ if we set $r_{-1} = 0$ and $s_{-1} = B$.

6.P.10

b) Now show, using induction and the above relations, that r_j, s_j are integers. (They are positive by problem 34).

37. Let $\sqrt{\frac{A}{B}} = (q_0, \overline{q_1, \ldots, q_m}) = (q_0, \overline{q_1, \ldots, q_{2m}})$ where m is odd. We know from problem 33 that $q_j = q_{2m-j}$, $j = 1, \ldots, 2m-1$. Show that $q_j = q_{m-j}$, $j = 1, \ldots, m-1$.

38. a) Show that if η is the second root of (a, b, c) then $-\frac{1}{\eta} = \frac{b + \sqrt{\Delta}}{c}$.

b) Find the period of $(-1, 2, 3)$ by calculating the complete quotients of $x_0 = \left| -\frac{1}{\eta} \right|$ and using them to construct $-\frac{1}{\eta_i}$ for each form f_i in the period, (this amounts only to alternating their signs), and reconstructing the forms from the $-\frac{1}{\eta_i}$. [This is probably the most efficient means of constructing a period.] Find the period again using y_0, and remember that the period now comes in reverse order.

c) Let $x_0 = (\overline{p_1, \ldots, p_{2n}}) = (p_1, p_2, \ldots, p_j, x_j)$ be a purely periodic continued fraction and let $y_0 = (\overline{p_{2n}, p_{2n-1}, \ldots, p_1}) = (q_1, q_2, \ldots, q_j, z_j)$, $q_j = p_{2n-j+1}$, $j = 1, \ldots, 2n$. Show that if

$$x_j = \frac{r_j + \sqrt{\Delta}}{s_j} \text{ then } z_j = \frac{r_{2n-j} + \sqrt{\Delta}}{s_{2n-j-1}}, \quad j = 1, \ldots, 2n,$$

where s_{-1} is set equal to s_{2n-1} and $z_0 = y_0$. Hint: the notation has been chosen consistent with the proofs of propositions 2 and 3 of §65. Choose an f_0 with first root $\omega_0 = y_0$ and form its cycle. Then $x_j = (-1)^{j+1}/\eta_j$ and $y_i = (-1)^i \omega_{2n-i}$. These facts and problem 36 lead to the desired relation.

d) Show that if x_0 has an odd period of length n then the above relation holds when n replaces $2n$ and $1 \leq i, j \leq n$. Here set $s_{-1} = s_{n-1}$.

6.P.11

39. Assume that $m \geq 2$.

a) Show that if $\frac{\sqrt{\Delta}}{B} = (q_0, \overline{q_1, \cdots, q_m})$ and if we set $s_{-1} = B$ then $r_{m-i-1} = r_i$
and $s_{m-i-2} = s_i$ for $i = 0, \ldots, m-1$. Hint: Set $(q_m, q_{m-1}, \ldots, q_1) =$
$(p_1, p_2, \ldots, p_m) = (p_1, p_2, \ldots, p_j, \pi_j)$, $1 \leq j \leq m$. Then determine
π_j in terms of r_k, s_k, by problem 38b. Next show that
$(\overline{p_1, p_2, \ldots, p_m}) = q_0 + \frac{\sqrt{\Delta}}{B}$

which allows a second determination of π_j in terms of r_ℓ, s_ℓ.
Equating the two expressions gives the required relation.

b) Observe the r_i, s_i for $\sqrt{\frac{13}{7}}$ (see problem 35). Note that the r_i are
periodic in a similar way to the q_i, but shifted. Note that the s_i
are periodic in a slightly different sense, since the periods share
the first and last terms if we require them to be symmetric.

40. Assume $m \geq 2$.

a) Show that if $r_i = r_{i+1}$ for some i with $0 \leq i \leq m-2$ then m is even
and $i = (m-2)/2$. Hint: Show that ω_i and ω_{m-i-2} are identical,
which, since m is the shortest period, implies $i = m-i-2$.

b) Show that if $s_i = s_{i+1}$ for some i, $0 \leq i \leq m-2$, then m is odd and
$i = (m-3)/2$.

c) Show that the converse of the a) and b) hold:
If m is even and $i = (m-2)/2$ then $r_i = r_{i+1}$.
If m is odd and $i = (m-3)/2$ then $s_i = s_{i-1}$.

d) Observe that, for the calculation of q_i, r_i, s_i of $\sqrt{\frac{A}{B}}$, it is only
necessary to calculate until the situation of a) or b) occurs. The
remainder of the q_i, r_i, s_i are then completely determined.

6.P.12

The following problems treat the case of $\sqrt{\Delta}$ where Δ is a positive integer not the square of an integer. Notation is as for the previous group of problems.

41. a) Show that if $s_i = 1$ then $i+1 = km$. Hint: Show that $\omega_{i+1} = \omega_0$.

  b) Show that $s_{m-1} = 1$.

42. Use problem 34 to show that $r_j < q_0$ and $s_j < 2q_0$.

43. Show that if $j \neq km$ then $q_j < q_0$. Hint: show $s_{j-1} \geqslant 2$, for $j \neq km$.

44. Show that if $\sqrt{\Delta} = (q_0, q_1, \ldots, q_j, \omega_j)$ where $q_j = 2q_0$ and $q_k < 2q_0$ for $k < j$ then $j = m$ and $\sqrt{\Delta} = (q_0, \overline{q_1, \ldots, q_j})$. Observe thus that the appearance of $q_j = 2q_0$ signals the end of the period.

45. Show that $\sqrt{k^2+1} = (k, \overline{2k})$.

The following problems 46-48 address the efficient solution of $ax^2+2bxy+cy^2=m$, where $\Delta > 0$. We define a semiperiod of forms to be those forms in a period with $a < 0$ or, equivalently, those forms with positive first root. We recall that a form may be instantly reconstructed from its first root.

46 a) Show that the semiperiod of f may be obtained by expanding the first root ω_0 of f in a continued fraction*. Moreover, show that the semiperiod becomes known as soon as some pair of final denominators ω_{2j} and ω_{2j+2K} are identical. Hint: Observe that reduced quadratic irrationals correspond to reduced forms and have purely periodic expansions. Also note that the first roots of forms in the period

*The first root ω_0 is often negative so that the first partial quotient q_0 is negative. This does not effect the algorithms.

6.P.13

    are all equivalent, and make use of the methods in the proof of proposition 3 of §65.

b)   Find the semiperiod of the forms in problems 3d), e).

c)   Find a transformation T which transforms the original form into a form of the semiperiod. Hint: use the partial quotients found in part a) to construct the T for $\omega_0 \sim \omega_{2j}$.

d)   Find the period of the form of problem 3d by selecting any form in the period and forming it's right adjacent forms. Note the order in comparison to b). Hint: the k's for the right adjacent forms (see §52) are already available from the calculations done in b). They are the partial quotients of the first root of the selected reduced form, but in reversed order and with alternating sign, (see §65). These allow rapid computation of the b', and a" are then easily computed from the discriminant Δ.

47.   Now find one solution for each of the auxiliary forms found in problems 3d), e), f) for which a solution exists, by finding the semiperiods of each of these auxiliary forms, comparing them with the semiperiods found in problem 44b), and determine the transformation S which transforms the original form into the auxiliary form.

48.   Finally, using the methods in §67, find transformations S, which transform the original forms of 3d), e), f) into themselves. Use these S to give formulas for all solutions of the problems among 3d), e), f) which are solvable.

Bibliography

The following bibliography has been added by the translator for the benefit of those persons who wish to continue their studies in Number Theory and related disciplines. Brief discriptions have been included to indicate the scope of the books and the level of preparation expected of the reader. Books have been included which would naturally complement the present book or which extend its investigations, especially in those directions that interested Hurwitz. Natural sequels to this book would be [11,21,10] on algebraic number theory, [19] on quadratic forms, or [20] on continued fractions. Reference [11] touches briefly on the connection between an integral basis for an ideal and a quadratic form, which is important in bridging the gap between the present work and algebraic number theory.

Persons who are interested in following the current literature in number theory will want to become acquainted with the following topics, and I have included introductory books on these topics: Topological groups and their associated integrals [12], algebraic geometry [26] and elliptic functions [14].

All references are cited in the original language. Translations, when I am aware of them, are indicated in parenthesis following the citation; [F=French, G=German, E=English, R=Russian].

1. Artin, Emil, ALGEBRAIC NUMBERS AND ALGEBRAIC FUNCTIONS, Gordon & Breach, New York, 1967.

> Notes of lectures by Artin on algebraic number theory and the closely related area of algebraic functions. Treatment is by divisors (classes of valuations). Part II is a treatment of local classfield theory, with a very concrete treatment of the cohomology of groups. The book occasionally requires the reader to fill in details, but Artin's treatment will well reward these efforts.

2. Borevich, Z. I. & Shafarevich, I. R., TEORIA CHISEL, (THE THEORY OF NUMBERS), Moscow, 1972, (G,E).

> Excellent and very readable introduction to wide areas of number theory, including quadratic forms, algebraic number theory (using divisors) and analytic methods. Many excellent problems.
> English edition: NUMBER THEORY, Academic Press, 1966.

3. Cassels, J. W. S. & Fröhlich, A. (Editors), ALGEBRAIC NUMBER THEORY, Washington, 1967.

> Treatment of class field theory using group cohomology, with each chapter by a different author. Rather difficult reading, with some gaps that must be filled in by supplemental reading, but rewarding. Contains the first printed version of J. Tate's thesis (1950), which has played an important role in the modern formulation of class field theory.

4. Chandrasekharan, K., INTRODUCTION TO ANALYTIC NUMBER THEORY, Springer Berlin, 1968.

> A very readable little book packing an incredible amount of information into its 130 pages. Topics include the prime number theorem, Dirichlet's theorem on primes in an arithmetic progression, Weyl's theorems on uniform distribution, representations of numbers as sums of squares, and Hurwitz's theorem on rational approximation of irrationals. The reader is expected to be familiar with Riemann integration and complex analysis up through Cauchy's Integral theorem. The book has a sequel entitled ARITHMETIC FUNCTIONS.

5. Dirichlet, P. G. L. - Dedekind, R., VORLESUNGEN ÜBER ZAHLENTHEORIE, Braunschweig, 1863-93. Reprint, New York, 1968.

> Very readable introduction to number theory with a large amount of material on quadratic forms in a similar style (reduction of forms) to the treatment in the present book. The famous eleventh supplement is one of the earliest systematic treatments of algebraic number theory.

6. Euler, Leonhard, INTRODUCTIO IN ANALYSIN INFINITORUM, Lausanne, 1748,
German reprint Springer, Berlin, 1983 (G,F).

    This pioneering work on the foundations of analysis is still
    perfectly readable, and contains much number theory, especially
    continued fractions and the theory of partitions, the latter
    presented here for the first time. Partition theory concerns the
    number of ways a number may be written as a sum of various sorts of
    numbers, for example, odd numbers. This book is highly recommended
    for persons wishing a historical perspective of number theory.

7. Hardy, G. H. & Wright, E. M., AN INTRODUCTION TO THE THEORY OF NUMBERS,
Oxford, 1938, (many reprints), (G).

    Classic introduction to number theory in the English speaking
    world, with a very rich selection of topics. Pleasant reading with a
    leisurely pace and historical notes. Topics include prime numbers,
    representation of numbers by decimals, continued fractions,
    approximation of irrationals by rationals, Diophantine equations,
    quadratic algebraic number fields, arithmetical functions, partition
    theory, representation of a number by sums of squares or higher
    powers, geometry of numbers.

8. Hasse, Helmut, VORLESUNGEN ÜBER KLASSENKÖRPER - THEORIE, Würzburg, 1967.

    Very readable treatment of class field theory using analytic
    techniques and ideas introduced by Takagi (Strahlklassen). The first
    chapter contains an excellent treatment of the decomposition of
    primes under relative - Galois extensions of number fields. This
    book is an natural sequel to that of Hecke.

9. Hasse, Helmut VORLESUNGEN ÜBER ZAHLENTHEORIE, (2nd ed.), Springer,
Berlin, 1964,(G, E in preparation).

    Excellent introduction to basic and algebraic number theory,
    with explanation and motivation for the direction of the theory,
    including historical commentary. Fine treatment of Quadratic number
    fields and Gaussian sums. Prime factorization in Quadratic number
    fields is by means of an early form of divisors due to Kummer.
    English edition: NUMBER THEORY, Springer-Verlag, 1980.

10. Hasse, Helmut, ZAHLENTHEORIE, Academie Verlag, Berlin, 1969,(E).

    An encyclopediac treatment of elementary algebraic number
    theory, using divisors defined as equivalence classes of
    valuations. Designed as a handbook rather than a textbook, it is
    nevertheless readable though the style is much less informal than the
    VORLESUNGEN.

11. Hecke, Erich, VORLESUNGEN ÜBER DIE THEORIE DER ALGEBRAISCHEN ZAHLEN,
Leipzig, 1923 (E).

Unsurpassed definitive treatment of algebraic number theory in
its classical formulation, using ideals. This elegant and beautiful
treatment is essential reading for all students of algebraic number
theory who wish to gain a historical perspective, or who wish to see
a fairly complete treatment of the ideal-theoretic approach with a
minimum of abstraction.
English edition: LECTURES ON THE THEORY OF ALGEBRAIC NUMBERS, Springer-
Verlag, 1981.

12. Hewitt, Edwin and Ross, Kenneth, ABSTRACT HARMONIC ANALYSIS, Springer,
New York, Vol. I 1963, Vol. II 1970.

Introduction to analysis on topological groups, including the
necessary integration theory on a locally compact Hausdorff space.
Easy to read and useful as a reference when using modern adele-
oriented treatments of algebraic number theory.

13. Hurwitz, Adolf, MATHEMATISCHE WERKE (2nd ed.), Birkhäuser, Basel, 1964.

Hurwitz's papers are still a source of inspiration and any
student of number theory will find much of interest in them.
Hurwitz's habit of making his articles as self contained as possible
and his great gifts as an expositor make them very accessible.

14. Hurwitz, Adolf, and Courant, Richard, VORLESUNGEN ÜBER ALLGEMEINE
FUNKTIONENTHEORIE UND ELLIPTISCHE FUNKTIONEN, 4th edition, Springer, New York,
1964.

Elementary introduction to the theory of complex functions using
Weierstrass' approach through power series. The approach is very
algebraic. The second part deals with elliptic and theta functions,
and is among the most enjoyable introductions to their study. A
brief treatment of the elliptic modular function is included. The
third part, by R. Courant, emphasizes the more geometric aspects of
the theory, including the Riemann mapping theorem, Riemann surfaces,
and the Schwarz reflection principle. (A version of parts one and
two by Prof. Kritikos exists in Greek.)

15. Ireland, Kenneth, and Rosen, Michael, ELEMENTS OF NUMBER THEORY (2nd ed.),
Springer, New York, 1982.

A very readable introduction to many different areas of number
theory, with a continuing theme of finite fields running through
it. Contains many results and many varient approaches not obtainable
elsewhere in such an elementary form, including equations over finite
fields and the zeta functions of algebraic curves over finite
fields. The style is informal, there are many detailed historical
notes, and it has interesting problems and a fine bibliography.

16. Landau, Edmund, EINFÜHRUNG IN DIE ELEMENTARE UND ANALYTISCHE THEORIE DER ALGEBRAISCHEN ZAHLEN UND DER IDEALE, 1927, (Reprint, New York Chelsea 1949).

   A very efficient treatment of algebraic number theory (not really suitable for an introduction) which is then used to prove the prime number theorem for algebraic number fields. Necessary material on the zeta-function and on multidimensional theta-functions is developed in the book. Landau's style is not particularly chatty, but there is a great deal to be learned from this little book.

17. Neukirch, Jürgen, KLASSENKÖRPERTHEORIE, Mannheim, Wien, Zürich, 1969.

   Very readable introduction to the modern approach to class field theory using cohomology of groups. The necessary group cohomology is developed in the book, making it more self contained than most treatments using this approach.
English edition: CLASS FIELD THEORY, Springer-Verlag, 1985.

18. Niven, Ivan, DIOPHANTINE APPROXIMATION, Interscience-Wiley, New York, 1963.

   An easily read introduction to a topic not really touched on in the present text, this book is concerned with the question of measuring the accuracy with which an irrational number may be approximated by rational numbers under certain given conditions.

19. O'Meara, Timothy, INTRODUCTION TO QUADRATIC FORMS, Springer, New York, 1971.

   Easily readable treatment of quadratic forms using a completely different approach (the theory of algebras) from the present work. The first chapter is a pleasant introduction to the modern Adele-Idele treatment of algebraic number theory (without Haar measure). Factorization is by means of divisors, in the modern valuation theoretic form. Chapter V has an excellent introduction to the theory of algebras.

20. Perron, Oskar, DIE LEHRE VON DEN KETTENBRÜCHEN (2nd ed.), Leipsig, Teubner, 1929 (Reprint, New York, Chelsea, c. 1960).

   Encyclopediac textbook on continued fractions, containing most of what was known up to the time of printing. Quite easy to read, even for persons with a limited vocabulary of German. All the material on continued fractions developed in the present book in conjunction with quadratic forms is developed by Perron independently of quadratic forms.

21. Samuel, Pierre, THEORIE ALGEBRIQUE DES NOMBRES, Paris, Hermann, 1967, (E).

Very well organized and readable introduction to algebraic
number theory, using ideals for factorization. Introduction to
Dedekind domains. The last chapter treats the decomposition of prime
ideals under relative Galois extensions of a number field.
English edition: ALGEBRAIC THEORY OF NUMBERS, MacGraw-Hill, 1970.

22. Rademacher, Hans, TOPICS IN ANALYTIC NUMBER THEORY, Springer, Berlin,
1973.

Excellent introduction to analytic number theory, especially the
applications of special functions (zeta-function, theta-functions,
Eisenstein series) to number theory, and partition theory. Several
theorems on sums of squares are proved with theta-functions.
Rademachers exact partition formula for the number of ways a positive
integer may be written as a sum of positive integers is presented, as
well as more standard material like the prime number theorem.
Occasional sections are difficult, but may be read by those who
persevere. The sections of the book are largely independent of one
another.

23. Schoenberg, Bruno, ELLIPTIC MODULAR FUNCTIONS, Springer, New York, 1974.

A relatively elementary introduction to elliptic modular
functions with much emphasis on their number theoretic
applications. The reader should be familiar with the theory of
complex functionsup through the Riemann mapping theorem and the
Schwarz reflection principle.

24. Serre, Jean-Pierre, CORPS LOCAUX (2nd ed), Hermann, Paris, 1968, (E).

The now classical treatment of local class field theory from the
point of view of the cohomology of groups. The reader is referred to
other works for certain results on group coholomology in the third
section. Somewhat difficult to read but greatly repays the reader
for his efforts. Prerequisites: a standard knowledge of group theory
and the elements of algebraic number theory, for example [11, 21].
English edition: LOCAL FIELDS, Springer-Verlag, 1979.

25. Serre, Jean-Pierre, COURS D'ARITHMETIQUE, Paris, Presses Universitaires de
France, 1970, (E).

A very rich and very modern treatment of number theory,
including quadratic forms, p-adic numbers and the Hasse-Minkowski
theorem, sums of three squares, the Hilbert symbol and its product
formula, Dirichlet's theorem on primes in an arithmetic progression,
and Modular forms. Dispite the shortness of the book and richness of
the contents it is still quite readable. Chapter two, where p-adic
numbers are defined in terms of projective limits, is rather
abstract, but the reader should persevere through this single
difficult section.
English edition: A COURSE IN ARITHMETIC, Springer-Verlag, 1985.

26. Shafarevich, I. R., OSNOVY ALGEBRAICHESKOI GEOMETRII (FUNDAMENTALS OF
ALGEBRAIC GEOMETRY), Nauka, Moscow, 1972, (G,E).

        One of a number of recent introductions to algebraic geometry, I
cite this one because of its relatively gentle transition from the
classical to the modern approach. Prerequisite is a standard course
in modern algebra.
English edition: BASIC ALGEBRAIC GEOMETRY, Springer-Verlag, 1982.

27. Siegel, Carl Ludwig, TOPICS IN COMPLEX FUNCTION THEORY, Interscience-
Wiley, New York Vol. I 1969, Vol. II 1971, Vol. III 1973.

        The interplay between number theory and complex function theory
occasionally extends as far as the theory of Riemann surfaces and
modular functions of many variables. This work is an excellent
introduction to these topics, and can be read withamoderate degree of
perseverence.

28. Vinogradov, I. M., OSNOVY TEORII CHISEL (ELEMENTS OF THE THEORY OF
NUMBERS), Moscow, 1953, (E,F).

        This interesting book presents number theory in outline form,
and the student must then fill in the gaps by doing the problems.
Working through this book will be excellent training for research,
though only gifted students will be able to solve the problems
without referring to the hints, which bring the problems within reach
of the better students. Topics included are roughly Chapters one
through five of the present book.
English edition: ELEMENTS OF NUMBER THEORY, Dover Publications, 1954.

Index

Anti-Index. 96
Binary quadratic form, 157
Binomial congruence, 99
Class of integers mod m, 51
Common divisor, 15
Common multiple, 16
Complete system of least absolute
 residues mod m, 58
Complete system of residues mod m, 58
Composition of substitutions, 163
Composite number, 8
Congruence $a \equiv b \pmod{m}$, 51
Congruence class prime to m, 59
Congruence of degree k, 89
Congruence $x^2 \equiv a \pmod{m}$, 120-123
Congruence $x^2 \equiv a \pmod{2^k}$, 116-119
Congruence $x^2 \equiv a \pmod{p^r}$, 112-115
Continued fraction, 79-83, 213
Cosets of r mod m, 64
Criteria of divisibility
 by 2,5,3,9,11,7; 53-54
Day of the week formula, 54
Decomposition of a natural number
 into prime factors, 6,11
Degree of r mod m, 62
Descent, 2
Diophantine equation $x^n + y^n = z^n$, 29
Divisible by, 3
Division algorithm, 3
Divisor, 3
Divisor σ of a quadratic form, 198
Equation of Fermat (or Pell or
 Lagrange), 192-197, 201, 243
Equivalence class mod m, 51

Equivalence of irrational numbers, 219
Equivalence of quadratic forms, 170
Euclidean algorithm, 21-22
Euler's ϕ-function, 32
Exponent of the natural number r mod m, 62
Fermat's conjecture, 29
Final denominator of a continued
 fraction, 214
Form of degree r, 157
Greatest common divisor (G.C.D.), 15
Homogeneous integral function, 157
Identically congruent polynomials mod m, 89
Index, 96, 272
Integral function, 157
Jacobi symbol $\left(\frac{a}{b}\right)$, 139
Law of quadratic reciprocity 135, 137
Least absolute system of
 residues mod m, 58
Least common multiple, 16
Least non-negative system of residues mod m, 58
Least positive system of residues mod m, 58
Left adjacent quadratic form, 172
Legendre Symbol, $\left(\frac{a}{b}\right)$, 112
Linear Congruence, 68
Linear form, 157
Linear transformation of a quadratic form, 161
Module, 18
Natural number, 1
Negative quadratic form, 178
Non-Residue, Quadratic, 109
nth power residue mod m, 101
Paradivisor τ of a quadratic form, 198
Parallel unimodular substitutions, 169, 172
Partial fraction, 76-78

Partial quotients of a continued
  fraction, 214
Pell's equation 192-197, 201, 243
Perfect number, 13
Period of a quadratic form with
  $\Delta > 0$, 236
Polynomials identically congruent mod m, 89
Positive quadratic form, 178
Primary coset of r mod m, 64
Prime number, 6
Prime to one another natural
  numbers 15, 17, 23-25
Primitive Pythagorean triples, 26
Primitive quadratic form, 199
Primitive quadratic form of the
first type, 199
Primitive quadratic form of the
  second type,199
Primitive representation of an integer, 174
Primitive root for a prime number p, 96,
Principle of Descent, 2
Product of transformations, 163-164
Pythagorean triple, 26
Quadratic form, 157
Quadratic form, reduced, with
  $\Delta < 0$, 179
Quadratic form, reduced, with
 $\Delta > 0$, 226
Quadratic non-residue mod m, 109
Quadratic reciprocity, Law of, 137, 135
Quadratic residue mod m, 109
Reduced quadratic form with
  $\Delta < 0$, 179
Reduced quadratic form with
 $\Delta > 0$, 226
Reduced system of residues mod m, 59

Regular continued fraction 79-83, 213
Regular polygon with $2^m + 1$ sides
  constructible ruler and compass, 43
Relatively prime natural numbers, 15, 17, 23-25
Representation of a natural number as a
  product of primes, 8, 11
Representation of a number by an integral
  form, 157
Residue, 58
Residue, nth power, 101
Residue, Quadratic, 109
Residue, system of, 58
Residue, reduced system of, 59
Right adjacent quadratic form, 172
Roots of a quadratic form, 187
Substitution, 162
Symbol of Jacobi $\left(\frac{a}{b}\right)$, 139
Symbol of Legendre $\left(\frac{a}{b}\right)$, 112
System of linear congruences, 71
Twin prime, 43
Unimodular substitution, 168
Unimodular transformation, 168